Critical Mathematics Education

Critical Mathematics Education

Can Democratic Mathematics Education Survive under Neoliberal Regime?

By

Bülent Avcı

Foreword by

Antonia Darder

BRILL SENSE

LEIDEN | BOSTON

All chapters in this book have undergone peer review.

The Library of Congress Cataloging-in-Publication Data is available online at http://catalog.loc.gov

Typeface for the Latin, Greek, and Cyrillic scripts: "Brill". See and download: brill.com/brill-typeface.

ISBN 978-90-04-39021-8 (paperback)
ISBN 978-90-04-39022-5 (hardback)
ISBN 978-90-04-39023-2 (e-book)

Copyright 2019 by Koninklijke Brill NV, Leiden, The Netherlands.
Koninklijke Brill NV incorporates the imprints Brill, Brill Hes & De Graaf, Brill Nijhoff, Brill Rodopi, Brill Sense, Hotei Publishing, mentis Verlag, Verlag Ferdinand Schöningh and Wilhelm Fink Verlag.
All rights reserved. No part of this publication may be reproduced, translated, stored in a retrieval system, or transmitted in any form or by any means, electronic, mechanical, photocopying, recording or otherwise, without prior written permission from the publisher.
Authorization to photocopy items for internal or personal use is granted by Koninklijke Brill NV provided that the appropriate fees are paid directly to The Copyright Clearance Center, 222 Rosewood Drive, Suite 910, Danvers, MA 01923, USA. Fees are subject to change.

This book is printed on acid-free paper and produced in a sustainable manner.

Contents

Foreword: A Living Mathematics for Democracy VII
   Antonia Darder
Acknowledgements XIII

1 **Setting the Stage** 1
  Research Context 2
  Research Methodology 3
  End-of-Unit Projects (EUPs): A Response to Neoliberal Pedagogy 4
  Outline of the Book 5

2 **The Standardization Movement in Education** 8
  Reconnaissance Stage for Critical Participatory Action Research 8
  Student-Developed Multiple-Choice Tests: EUP 1 9
  Concluding Discussion of EUP 1 22
  Neoliberal Hegemony in Education 23

3 **Class Consciousness and Mathematical Literacy** 37
  Planning and Objectives 37
  Race, Class, and Critical Pedagogy 49
  Concluding Discussion of EUP 2 52

4 **Collaborative Versus Competitive Learning** 53
  Planning and Objectives 53
  Concluding Discussion of EUP 3 64

5 **Mathematical Inequality and Socioeconomic Inequality** 67
  Planning and Objectives 67
  Concluding Discussion of EUP 4 84

6 **Student Loan Crises** 88
  Planning and Objectives 88
  Concluding Discussion of EUP 5 103

7 **Critical Mathematics Education: A Bottom-up Response
to Neoliberal Hegemony** 107
  Critical Mathematics Education and Dialogic Pedagogy 107
  Critical Mathematics Education and Collaborative Learning 118
  Critical Mathematics Education and Inquiry-Based Learning 126

VI   CONTENTS

8 **Critical Mathematics Education and Citizenship**
   in the Neoliberal Era   134
   Democracy in the Classroom   134
   Democracy in a Form of Dialogue   136
   Mathematical Literacy and Citizenship   137
   Critical Pedagogy and Critical Citizenship   140
   Neoliberal Hegemony and Mathematics Education for
        Critical Citizenship   143

9 **Conclusions**   146
   Dialogue, Collaboration, and Inquiry   146
   The Mathematics Classroom as a Micro Society   147
   Citizenship and Mathematics Education   149
   Curricular Materials in CME   150
   Micromanagement and Control   151
   Participatory Action Research and Critical Mathematics Education   152
   Limitations and Suggestions   152
   Final Word   153

   References   155
   Index   164

Foreword: A Living Mathematics for Democracy

Antonia Darder

> Our lives are lived, in actual fact, among changing, varying realities, subject to the casual play of external necessities, and modifying themselves according to specific conditions within specific limits; and yet we act and strive and sacrifice ourselves and others by reference to fixed and isolated abstractions which cannot possibly be related either to one another or to any concrete facts. In this so-called age of technicians, the only battles we know how to fight are battles against windmills.
>
> SIMONE WEIL, *An Anthology* (2000, p. 223)

∴

Writing in the 1930s, Simone Weil's words presaged the growing alienation that would unfold as a consequence of advancing capitalism and its disregard for the dignity of our humanity. During the last four decades, the regime of neoliberalism has taken center stage in its political economic efforts to prevent the full democratization of society. Through a long-standing belief in the power of absolute measurement, the knowledge realm of mathematics has been, indeed, systematically privileged, as well as colonized as the handmaiden of capital, in the quest to produce a reified world view of binaries and standardizations that work to obstruct the majority of citizens from comprehending, let alone participating within, the structures of power that determine our lives.

Through the process of hegemonic schooling, citizens are individualized and depoliticized, effectively stripped of imagination and curiosity, as they are ushered into the world as technicians of the wealthy and powerful. Traditional mathematics education has been informed by a pedagogy blinded by an ever-narrowing rationality and approach transfixed by the myths of fixed and isolated abstractions. The monstrous ideology of neoliberalism, intent on absolute domination of power and wealth by a few, has declared itself victorious at the expense of our embodied and messy human existence—the same messy human existence from which genuine democratic life must emerge. In essence, the banking approach to education (Freire, 1971) prevalent in mathematics education is the perfect cover-up, in that it not only despises human messiness as irrational, it also blames those who are unable to penetrate the realm of mathematics for their own failure. Hence, the fundamentally unjust values of meritocracy, so endemic to the current neoliberal ethos of schooling, are reproduced, as inequalities are justified and perpetuated.

Reconceptualizing Mathematics Education

> Insofar as mathematics is about reality, it is not certain, and insofar as it is certain, it is not about reality. (Albert Einstein, 1922)

Albert Einstein, the consummate scientist of the twentieth century, was more apt than his colleagues of the time to question the absolute nature of mathematics, calling into question the certainty of its relationship to reality. This, of course, did not mean that Einstein did not possess a deep respect for mathematics, but rather than he seemed to comprehend that mathematics, stripped of experience and human life, could not in itself inform our understanding of the world. He recognized that mathematics, like language, is a product of the human mind, expressed as symbols that are created to represent life but are not life. It is a bit like seeing oneself in the mirror. The glass reflects an image of ourselves but in no way is that image the embodied us. Hence, any desire to engage mathematics education critically must begin within the realm of lived experience—the lived experiences of the teacher and the students. Only through such an examination is it possible to not only understand mathematics in a way where power is implicated and social inequalities exposed, but also to consider how critical mathematics education itself can serve as a possible space for the teaching and living of democratic life.

With this in mind, *Mathematics Education for Critical Citizenship* is an excellent example of the ways that our lived experience can serve as a powerful impetus and strategy for the construction of mathematics knowledge through naming the world, both as an individual researcher and as a collective participant with students/participants, without whom we could not genuinely come to know or appreciate our practice. As such, this volume provides a glimpse into an excellent pedagogical study, where both the author's life as a critical educator-researcher and his mathematics classroom are utilized as critical sites of inquiry. Through this interrogation, the meaning of citizenship is explored within an educational context where such questions are typically ignored, let alone given primacy.

By so doing, Bülent Avcı centers his study on a few central questions, which focus on the attainability of critical pedagogical practice within the constraints of neoliberal educational conditions, which are essentially in direct contradiction to his larger emancipatory vision of mathematics education.

FOREWORD: A LIVING MATHEMATICS FOR DEMOCRACY IX

Here, the underlying purpose of the standardization movement's so-called reform efforts are unveiled, as Avcı critiques the imposition of a one-size-fits-all approach that drives mathematics educational practices. At the heart of this deeply unjust and decontextualized neoliberal strategy is a conservative political drive to destabilize and undo the very meaning and practice of public education and undermine the right to education for all children. Hence, an education that was once (at least in theory) meant to encompass the liberal democratic ideal of citizenship and civic participation has been replaced by an authoritarian top-down tenet, where, by producing efficient and obedient workers, schools can function as economic engines for the nation. Accordingly, public school teachers have been converted into technicians who are expected to enact "best practices" tied to an "evidence-based" or scientific preset curriculum and time frame, indifferent to the lived histories and contemporary conditions faced by students, particularly students from working-class and racialized communities.

Avcı carefully explores the instrumentalizing restrictions imposed by neoliberal policies and their impact on the teaching of mathematics, in the hopes of articulating key dimensions related to his efforts to bring a critical mathematics education praxis to fruition, even within an educational context where the teacher must contend with the limit-situations, as Freire (2000) would say, and create counterhegemonic possibilities within an arena of hegemonic schooling. Toward this end, Avcı's reconceptualization of critical mathematics education, which emerges directly from his critical participatory engagement with students, offers an outstanding contribution to the literature in the field—a literature that has been steadily expanding over the last two decades.

Through the use of participatory nongraded activities, student-generated exams, and a critical mathematics literacy approach, students find classroom space to find their voices, create the conditions for their democratic participation, and build solidarity as they collaborate and learn together. By doing so in a pedagogically conscientious and consistent manner, teacher and students work as *revolutionary partners* (Freire, 2000) to challenge neoliberal classroom policies and practices rooted in competition, individuality, and instrumentalization, as they transform together the learning environment in the moment. This critical approach also offers students a powerful opportunity to experience embodied democratic participation in ways that not only enhance their understanding of mathematics within the classroom, but also out in their own world. Just as Avcı's students experienced greater opportunities to learn with one another when engaged through a critical mathematics literacy approach, so too was Avcı as their teacher transformed by the teaching-learning relationship with his students, as together they entered in a critical praxis that dialectically helped to shatter the school/world divide generally conserved by hegemonic schooling. Of this he asserts, "While improving their content knowledge, the students related their learning to a larger society and developed a bottom-up response to neoliberal educational changes. The project helped the students and me to engage in a structural analysis of neoliberal educational policies and implementations. In doing do, we developed and exercised critical mathematical literacy"

In reading *Mathematics Education for Critical Citizenship*, what is illustrated repeatedly is that the only effective manner to counter neoliberalism's mania with scientifically packaged curriculum and teacher accountability within the classroom is an unwavering commitment to a critical praxis. However, this requires that teachers engage critically and thoughtfully the conditions within schools, the lives of their students, and the larger social, political, and economic context that informs our lives. In this sense, a critical mathematics approach requires teachers who are prepared to counter the perverted market logic of neoliberalism, steeped in the language of school choice, high-stakes meritocracy, and student surveillance. Hence, critical mathematics as reconceived by Avcı in his study functions as a *bottom-up* response to the hegemonic conditions produced by the greed and indifference of neoliberal reform.

Certain important key points are made that point to the teacher's knowledge and the sensibilities that must encompass a critical praxis. The teacher, for example, must have an understanding of class consciousness and how schools reproduce class formation through traditional forms of mathematics education and the instrumentalization and fracturing of the math curriculum. In contrast, a teacher utilizing a critical mathematics approach places dialogue at the heart of the teacher's engagement with students. This means that the teacher does not expect students to abandon their lived experiences or who they are in the classroom, but rather sees these as sources of strength and significant foundations from which students enter into the process of critically learning mathematics, as well as their coming to understand the emancipatory potential of mathematics in the world. With all this in mind, teachers of critical mathematics, who are intent on challenging neoliberal practices in the classroom, must bring an emancipatory understanding of culture politics, economics, history, knowledge construction, ideology, critique, dialogue, and conscientization, as well as their excellent knowledge of mathematics.

Avcı's research methodology is worthy of greater consideration and replication, given his uncompromising emphasis on finding or creating spaces within a neoliberal mathematics classroom for critical engagement between teacher and students. His methodology is, indeed, consistent with a critical pedagogical approach to mathematics education in that the themes of his inquiry were anchored in his students' life-world; the inquiry process welcomed ideas that might be considered unpopular within the mainstream curriculum; the participatory approach was facilitated with the students in order to enable their critical engagement and participation; and the process of reflection and self-evaluation was an effective alternative to traditional banking education. Most importantly, the participatory approach supported the evolution of the mathematics classroom as a genuine community, where relationships were based on horizontal dynamics, cooperation, collaboration, and mutual respect—all key critical principles of democratic communities.

Making the Pedagogical More Political

As critical teachers working in schools, we can make the pedagogical more political….One approach might be to organize a radical pedagogy of citizenship around a theory of critical literacy. (Henry Giroux, 2005)

In *Schooling and the Struggle for Public Life*, Giroux posits important challenges related to the importance of teachers using their knowledge and insights to thoughtfully redefine the terrain of politics and citizenship within the classroom, as also part of a wider collective struggle out in the world. The sense of teachers making the *pedagogical more political* and organizing their pedagogy around issues of citizenship and critical literacy is at the heart of Bülent Avcı's study, where he rightly connects his own critical mathematics praxis to issues of citizenship. Democracy here is directly linked to important pedagogical and political questions about dialogue, citizenship, mathematics literacy, neoliberal ideology, political discourses, and students' development of critical thought.

Drawing on the writings of Freire and Habermas, Avcı asserts the power of critical citizenship to create spaces for individuals and communities to participate in decisions related to their own lives, despite neoliberal constrictions, through forms of political activism that are informed by a public pedagogy. Here, space is created for dialogue, where there is commitment to remain at the table, despite differences and struggles that may ensue. Critical citizenship also requires civic courage to speak the unspeakable and to listen to what might seem the unlistenable, in an effort to create common ground for work together. At the center of the practice of democratic citizenship must be both a healthy skepticism in order to critically question commonsensical notions, as well as working together to translate our common knowledge into social action for transformation. This also points to the manner in which critical literacy can assist teachers and students discern the changes that occur within the social, economic, and political landscape, so that we can respond together more clearly to the conditions we face within schools and communities.

Throughout this excellent book, Avcı provides well-researched and well-written discussions and conclusions that counter the helplessness often experienced by teachers who feel trapped in the colonizing politics and practice of neoliberal life in schools. This can be particularly the case with mathematics education, given its conservative foundational epistemology. Nevertheless, what is heartwarming and encouraging is the manner in which critical mathematics education, when implemented through praxis of reflection, dialogue, and action, resulted in the promotion of critical citizenship and expressions of democratic life within the classroom that encouraged students to embrace democratic values in their relationships within the classroom and beyond.

Avcı provides a powerful conclusion to this study—the first to undertake a participatory critical inquiry that so eloquently connects mathematics education to the struggle for democracy and critical citizenship. Underlying this excellent treatise are critical moral questions related to our responsibility as educators to genuinely *make the pedagogical more political* if we are indeed committed to a future where mathematics education is no longer the tyrannical boogeyman used to sort and sift away working-class and racialized populations from opportunities within schools and society. Through his study, Bülent Avcı carves a path toward a living mathematics for democracy—a critical mathematics approach fully committed to the making of critical citizens with the knowledge and skills to both participate in and transform their world.

References

Einstein, A. (1922). *Geometry and experience*. London: Methuen & Co. Ltd. Retrieved from http://www-history.mcs.st-andrews.ac.uk/Extras/Einstein_geometry.html
Freire, P. (1971). *Pedagogy of the oppressed*. New York, NY: Seabury Press.
Freire, P. (2000). *Pedagogy of the oppressed*. New York, NY: Continuum.
Giroux, H. (2005). *Schooling and the struggle for public life*. Boulder, CO: Paradigm Press.
Weil, S. (2000). The power of words. In S. Miles (Ed.), *Simone Weil: An anthology*. New York, NY: Grove Press.

Acknowledgements

I would like to thank my wife, Samantha, and my daughter, Leyla: This book would not have been possible without their support. I would also thank all of my students who took part in this participatory action research project.

I am thankful to Antonia Darder and Kemal Inal for their help and solidarity that widened my horizons and supported my writing during this book project. A special thanks to Ole Skovsmose who shared his ideas and work with me throughout this project. Special appreciation goes to Douglas Vipond for his editorial suggestions and my colleague Jamie Snowden for reading the chapters. Finally, I would like to thank John Bennett and Jolanda Karada from Brill Sense for their support with this book.

CHAPTER 1

Setting the Stage

This book is based on a high school math teacher's PhD dissertation that investigates the scope and limitations of critical mathematics education in spite of neoliberal changes in education. Therefore it can be read as professional-personal journey of a high school math teacher in U.S.—a teacher who was frustrated with the top-down, imposed corporate agenda in education to turn public schools into testing factories and sites for profit; an agenda that pictures teachers as salespersons and students as passive customers; an agenda that sees students as nothing more than the future workforce, ignoring their needs as humans and citizens, and thus reducing K-12 education to a course of vocational training.

In my journey of seeking alternative pedagogic approaches to the top-down, imposed, corporate-driven changes in education, I have been inspired by critical (educational) theorists, especially by Jurgen Habermas's theory of "ideal speech situation" and Paulo Freire's dialogic pedagogy. These approaches are complementary; both are aimed at achieving non-dominating (democratic) communication. Exposing myself to the literature in critical pedagogy and critical mathematics education enabled me to see the big picture.

Recent theoretical studies in critical pedagogy (CP) and critical mathematics education (CME), and given a dearth of classroom-based research, generate two main conclusions: First, with its top-down, imposed policies and implementations, market-driven educational changes curtail the potential of educational practice to promote democratic values and critical citizenship. Second, it is fundamentally important for students and teachers to be engaged with an educational practice that enacts humanizing dialogic education and promotes participatory democracy and critical citizenship. This situation led me to ask myself a major question:

> *Can critical pedagogy be an attainable practice within educational conditions that are contradictory to the larger emancipatory vision of mathematics education?*

Through a PhD program, I conducted a critical participatory action research (CPAR) project with the students in my own class to answer the question. From this point of departure, the book offers classroom-based data to reconcile these contradictory stances and investigates the ways in which critical mathematics

© KONINKLIJKE BRILL NV, LEIDEN, 2019 | DOI:10.1163/9789004390232_001

education could respond to the tension between the needs of a neoliberal system and the needs of individual students to fulfill their potential as human beings and citizens.

Drawing on rich ethnographic data, the book responds to ongoing discussions on the standardization of the curriculum. I reconceptualize CME by arguing that despite obstructive implications of market-driven changes in education, a practice of CME to promote critical citizenship could be implemented through open-ended projects that resonate with inquiry-based collaborative learning and dialogic pedagogy. Neoliberal hegemony can thereby be countered. However, this counterhegemonic practice necessitates transforming the classroom into a community of learners to democratize classroom life and create opportunities to promote participatory and social justice-based citizenship.

In this book, I identify certain limitations of CME resulting from (a) being a counterhegemonic practice enacted within an educational (neoliberal) system while criticizing the same system, and (b) a lack of hands-on learning materials and professional supports to enact a CME program. The book suggests pedagogic and curricular strategies for mathematics teachers to cope with these obstacles.

Research Context

I conducted the research in a high school in the state of Washington, USA, where I have been a mathematics teacher since 2006. In 2014–2015, the school had 1,942 students. Distribution of students' ethnicity was as follows: White 45%, Asian 17%, Hispanic 13.5%, African American 12.6%, Native American 1.4%, multi-racial 10.6%. Students came mostly from poor families: In the year of the research, 45% of students received free or reduced-cost meals as they met income eligibility requirements set by the federal government.

Students have to pass a number of standardized tests as a graduation requirement, some of which have changed several times. Under the federal No Child Left Behind (NCLB) Act, the Washington Assessment of Student Learning was administered between 1997 and 2009, replaced in 2010 by the High School Proficiency Exam (HSPE). High school mathematics end-of-course (EOC) exams replaced the HSPE in 2011. Since then, the EOC exam has been the state-level standardized test that students must pass. From 1997 to 2014, the success rate of students in my high school fluctuated between 40% and 60%. Common core curriculum standards were adopted in Washington in 2014; from the 2015–2016 school year on, the Common Core Assessment Test (CCAT), a fully computerized test,[1] replaced EOC exams. In addition to these, as part of college admission requirements, some students take exams such as Scholastic Assessment Test (SAT) and course-based advanced placement exams administered by a private company, the College Board.[2] In December 2015, the Obama government revised the NCLB Act that limits involvement of the federal government in educational decisions.[3]

The mathematics course (Advanced Algebra) from which I collected data September 5, 2014, to May 25, 2015, was scheduled from 9:00 a.m. until 10:30 a.m. during second period every day throughout the 2014–2015 school year. There were 28 students, aged 14 to 17. In the classroom, I had instant access to the Internet; an electronic projector was connected to my tablet PC and document camera. I purposefully ordered tables for my classroom. In a traditional classroom, students sit facing the board or the teacher's table, as the teacher is regarded as the ultimate source of knowledge and therefore an authority figure. Instead, inspired by critical pedagogy, I had six groups: each group included four tables with chairs in order to create a classroom ambiance conducive to non-authoritarian and dialogic teacher-student and peer relations.

Given neoliberal changes in education, raising test scores became the main point of the standardization movement, and it has resulted in similar teaching practices and classroom routines in mathematics classrooms nationwide, including my school. From a market-driven perspective, this approach is called "best practice": teachers are instructed to follow the same routine and focus on similar test-preparation activities. A typical lesson begins by reviewing the previous day's homework or presenting warm-up problems on the board (to make the transition to a new topic or to review topics that have already been covered). The teacher then introduces a new subject based on the standardized sequence; hands out a worksheet that includes repetitive mechanical exercises on a related topic, and has students practice them; before the end of class, the teacher assigns homework for the next day.

As I would like to consider my professional practice as praxis, within this typical daily classroom routine I strove to apply critical mathematics education (CME) in my class as much as possible. This point precisely frames the main question that the book offers an answer to: Given the restrictions imposed by standardized education, how much space can be cleared to practice CME?

Research Methodology

For classroom-based research in education, it is crucial that the methodology fits the dynamics of daily teaching practices. Therefore I preferred critical participatory action research (CPAR) as it resonates well with critical pedagogy

4   CHAPTER 1

and daily classroom teaching practices. Action research, according to W. Carr and Kemmis (1986), is

> a form of self-reflective enquiry undertaken by participants (teachers, students or principals, for example) in social (including education) situations in order to improve the rationality and justice of (a) their own social or educational practice, (b) their understanding of these practices, and (c) the situations (and institutions) in which their practices are carried out. (p. 162)

Considering the aims of CME as a subset of critical pedagogy, which guided my classroom teaching, CPAR conceptually resonates with my research project for three reasons. First, CPAR itself is a learning process that performs inquiry *with* participants (students), not *on* them; it does not objectify participants. On the contrary, it embraces each participant as an equal contributor. Similarly, a critical stance to mathematics education oriented toward critical citizenship envisions students as equal partners in the knowledge creation process and embraces a horizontal student-teacher relationship (Freire, 2000).

Second, CPAR is emancipatory. It is a process in which people explore root causes of oppression in order to break free of these constraints. Similarly, CME is critical of dominant (neoliberal) education systems; it envisions a mathematics education that empowers students. Cycles of plan-act-observe-reflect in CPAR resonate with daily teaching practices and provide opportunities to unpack oppressive dynamics and eliminate them.

Third, CPAR puts emphasis on engaging "in communicative action [based on dialogue]" with research participants to reach "intersubjective agreement" (Kemmis, McTaggart, & Nixon, 2014, p. 68). CPAR opens "communicative spaces that permit and foster" collective learning and doing (Kemmis, 2008, p. 126). Similarly, a dialogic teacher-student relationship is a focal point of CME; it therefore takes a central place in my research project. Freire (2000) argues that humanizing education necessitates a dialogical learning-teaching process.

End-of-Unit Projects (EUPs): A Response to Neoliberal Pedagogy

As shown in Table 1.1, end-of-unit projects (EUPs) are lessons over two consecutive days in 90-minute block periods, one period each day. I designed each EUP as to be an inquiry-based collaborative lesson to promote dialogic teaching and learning. During each EUP, I collected data from students' journals, their work, whole-class discussions, field notes, and my reflective

SETTING THE STAGE

TABLE 1.1 Content and themes of end-of-unit projects (EUPs)

EUP	Mathematics content	Theme
1	Linear equations and functions	Neoliberal changes in education and standardized assessment
2	Multipart functions: analysis of domain and range	Critical mathematical literacy: Class consciousness
3	History of mathematics	Collaboration versus cooperation
4	Systems of inequality	Community service: Solidarity versus charity
5	Exponential functions	Student loan debt crisis: Dialogical teaching of mathematics

journal. I positioned each EUP as a plan-act-observe-reflect cycle. I inscribed each EUP in the mandated curriculum. In order to meet the school district's requirement, the math content of each EUP was determined by the standardized curriculum. However, the theme of each EUP was determined by me. For example, the content of EUP 1 was linear equations and functions, but its theme was neoliberal changes in education and assessment.

Each EUP began with an introduction. Then students worked on the EUP in small groups. This was followed by a whole-class discussion to reflect on our learning experience and the theme of the EUP. At the end of each EUP, students made entries in their reflective journals. Although each EUP had its own content and theme, dialogic pedagogy and inquiry-based collaborative learning were central to all EUPs. Between any two EUPs, there was about 3–4 weeks classroom time. During this time, I followed the district-mandated scripted curriculum to meet the needs of standardized education. A typical lesson started by checking the previous day's homework. Then I introduced the topic of the day and showed some examples. Then I handed out worksheets, and students worked on these repetitive exercises.

Outline of the Book

This book comprises nine chapters. In Chapter 2, I discuss curricular standardization and neoliberal changes in education in the context of EUP 1. In EUP 1, students designed multiple-choice math questions and reflected on whether these kinds of questions accurately assess their skills and knowledge.

In this context, I argue that standardized assessment cannot adequately measure students' knowledge and skills. Market-driven changes in education aim to turn schools into sites for profit, reduce education to training, and make teachers into trainers. I conclude that corporate reforms in education fulfilled none of its stated promises. Instead, they created a new sector called "edu-business," from which multinational textbook companies and education consultancy companies greatly benefited.

Chapter 3, featuring EUP 2, focuses on the concept of class-based society. I argue that class consciousness is an important identity marker that has been long forgotten under the neoliberal regime. Class consciousness is an essential element in an individual's life-world in order to be a critical citizen. Critical mathematical literacy plays a significant role in developing this class consciousness. Class and race are inseparable in the U.S., and therefore anti-racist movements need to be reconsidered as ways to counter neoliberal hegemony instead of supporting it.

The main theme of Chapter 4 is collaboration and competition in context of history of mathematics. Unlike the neoliberal argument, I suggest that collaboration has been more beneficial for humanity than competition. After discussing obstacles to the collaborative learning of mathematics, I conclude that if the learning process is to counter the hegemony of neoliberal pedagogy, it should be designed as collaborative.

Chapter 5 is based on EUP 4, which problematizes generosity, charity, and solidarity. It argues that critical educators should exercise caution when they organize community volunteer projects, as there is a fine line between charity and solidarity.

Chapter 6 illustrates that despite the fact that authoritative teaching is dominant in math education, dialogic pedagogy is attainable, provided that the classroom is organized as a community of learners.

Chapter 7 presents a reconceptualization of CME as a counterhegemonic practice of teaching and learning mathematics. I argue that CME can be a subversive educational practice through inquiry-based collaborative learning and dialogic pedagogy. Chapter 8 investigates the relationships among CME, democracy, and critical citizenship.

Chapter 9 presents the overall conclusion of the action research study. It explains that CME is not welcome in schools colonized by neoliberal pedagogy. A math teacher who wishes to practice CME must be an artist in order to apply CME while still meeting requirements imposed by the neoliberal system. However, it also shows that neoliberal hegemony can be countered by creating small openings in the class where students can experience democracy through dialogic pedagogy.

Notes

1 This new test is called Smart Balance Assessment Consortium (SBAC), which is a private company: their official website is: http://www.smarterbalanced.org/2013/06/practice-tests-now-available/. Even from a neoliberal perspective, the quality and cost of the SBAC is highly controversial: http://www.livingindialogue.com/strange-history-common-core-sbac-test-monster-adopted-washington-state/

2 Detailed information about the College Board and their testing business can be obtained at https://www.collegeboard.org/about

3 Further information can be obtained at http://www.nytimes.com/2015/12/11/us/politics/president-obama-signs-into-law-a-rewrite-of-no-child-left-behind.html?_r=0

CHAPTER 2

The Standardization Movement in Education

In this chapter, I present the preparation or reconnaissance stage of the action research project that occurred in the classroom. The chapter illustrates a way of providing students with opportunities to reflect on standardized assessments in mathematics. The two-day lessons I taught involve small-group work, whole-class discussions, student journals, and my own reflections on standardized assessment and collaborative learning.

In addition, I discuss the ideological and political background of market-driven changes and corporate agendas in education. The chapter provides an analysis of impacts of market-driven changes on curriculum, assessment, teaching, teachers, students, citizenship, and democracy.

Reconnaissance Stage for Critical Participatory Action Research

This stage formed the pilot to the development of end-of-unit projects (EUP). By the very nature of critical participatory action research (CPAR), students were co-constructers of knowledge, so their participation was crucial to the success of the project—this was research *with* students and not *on* students. Therefore, during the first five weeks of the school year, within the mandated curriculum, I integrated discussions and learning activities to articulate to the students the purpose of the study and the scope and importance of their voluntary participation.

I informed students that the activities and assignments would not be graded in the usual fashion; instead, the purpose of the lessons was to experience the enjoyment of learning mathematics in a new way. For instance, in this preliminary stage, I developed and facilitated two block lessons, conducted over two days, on the concept of slope and tangent of a given line and angle using real-life environments; the school is a two-story building with stairs in multiple locations. The students spent most of the first 90-minute period out of the classroom in small groups, with cameras and measurement tools. They calculated the slopes of the stairs and took pictures. The next day, groups presented their work along with calculations and pictures. One group concluded that the slope of the stairs next to the cafeteria was too steep and should be fractioned into two or three pieces to lower its slope.

My initial plan was to use these experiences to shape the classroom into a public sphere where students could easily exchange ideas and where a

© KONINKLIJKE BRILL NV, LEIDEN, 2019 | DOI:10.1163/9789004390232_002

THE STANDARDIZATION MOVEMENT IN EDUCATION 9

community of collective learning could be established. In this sense, the two-day lesson was a greater success than I had anticipated. I initially thought that inviting students to participate in learning activities that were not conventionally assessed (such as awarding a grade of A or B) would be a major challenge to be addressed in the reconnaissance stage. During the two-day lesson, I asked my students whether they would have been more interested if the activity was graded and recorded in the grade book. In their feedback, students confidently stated that they enjoyed the activity, and being graded would not have made any difference to their participation level. Therefore, I concluded that my initial anticipation was not supported.

Subsequently, I designed each EUP as a non-graded activity. Reflecting on the students' positive reaction to ungraded learning activities was crucial to developing my professional understanding about participation and engagement. I inferred from this pilot lesson that if a learning activity is joyful for students, they actively and enthusiastically participate, whether graded or not. Conversely, I concluded that if a learning activity or lesson was considered boring by students, they would usually participate—just because it was graded. At that point, I resolved that for this project all EUPs should be related to students' life-world and interests in order to maintain their active participation and motivation.

Student-Developed Multiple-Choice Tests: EUP 1

Planning and Objectives
This was the first project aimed at creating a communicative space as a concrete step to molding the class into a democratic public sphere where shared concerns and problems about our education can be negotiated. The purpose of creating such a space, as outlined by Kemmis et al. (2014), is that students "need both space and permission to bounce ideas around" (p. 91). If students felt uncomfortable expressing their thoughts, this study would not have qualified as CPAR.

I determined standardized assessment in mathematics as the theme of this project for several interconnected reasons. Multiple-choice (MC) tests take up considerable time in the students' life-world. They are directly relevant to students' lives; therefore, every student in the class could easily participate in group work, classroom discussions, and peer dialogues. In other words, standardized assessment can be considered a shared concern of students and teachers. Three primary objectives for this EUP were:
- Improving students' conceptual knowledge and practical skills with respect to mechanical aspects of related content (linear equations/functions and

graphs): common misconceptions, procedural mistakes, and other elements that often lead to wrong answers on MC questions.

- Providing students with an opportunity to build a collaborative learning environment where dialogue is one of the forms of learning: opening a communicative space.
- Helping students gain insight into the social, political, and economic implications of standardized assessment to connect life in the classroom to the larger sociopolitical system and critically reflect on it, thereby countering neoliberal hegemony.

As I inscribed this project in a mandatory unit that covers linear functions and equations, I chose the mathematics content of the project to be linear functions and equations. I first scaffolded developing an MC question in mathematics on the board; students asked questions and provided feedback in this process. Students created their own groups, and I provided each group with a (non-MC) question. Then each group produced the options for their own MC question. Once each group completed designing their choices, they presented their questions and explained their process. Afterwards, each group conducted a brief search on the web to collate a range of ideas for and against using MC tests.

Day 1

I had planned to have a whole-class discussion on the topics of collaborative learning, inquiry, and dialogue; however, I changed my mind after realizing that letting students experience the project first and having the discussion at the end would be more productive, as they would have a concrete experience to reflect on. Having briefly explained the purpose of the project, I scaffolded the process of writing an MC question about linear equations.

After the scaffolding stage, I let students form groups of three or four students. I then handed out questions I'd written earlier—one question per group (see Appendix A for a sample of the students' work on the project). As students requested assistance, I answered their questions and provided suggestions. For example, Nick raised his hand and said, "How come we can't put wrong random answers, how come we have to calculate?" My response was, "If you put random answers, common mistakes would probably never lead to one of the choices....In that case, you would probably check your solution process one more time; you would be less likely to fall into the trap and end up with the wrong answer." Nick answered, "It makes sense, I never thought of it that way." Nick could have addressed the same question to his peers first. But he preferred to ask the teacher. In a traditional classroom, the teacher is usually regarded as the ultimate source of knowledge. My observation prompted me to

THE STANDARDIZATION MOVEMENT IN EDUCATION

note in my reflective journal that it would take time to shift from a traditional teacher-centered classroom to a facilitative one.

The group work progressed quite smoothly overall. In my observation, peer discussions mostly focused on the most common mistakes for each stage in the solution manual. As I circulated among the groups, I noticed that mechanical questions seemed easy for some students. However, other students were improving their understanding of the mathematical concepts. For example, at one point, Cincere asked his group, "You put it in slope-intercept form...why is that?" Jennifer responded, "It's easier to graph slope-intercept form." I closely observed another group as they were debating the order of operations— whether -1^2 is same as $(-1)^2$. In this instance, a group member explained to his peers why these two expressions were not the same.

Once each group completed their MC question and were ready to share their work, two groups volunteered to present their work; there was not enough time for all eight groups to present. I urged the class to ask questions of their peers who were sharing their work, but this moment did not trigger a lively discussion of the presentations. It might be because, as 10th and 11th grade students, they had already formed an attitude that does not include space for questioning, which resonates with the "transmit-receive" paradigm. We may need to work on skills for questioning.

Day 2

I briefly reminded my students what we had done yesterday and gave back each group's paperwork. Before the whole-class discussion, each group did a brief web search on standardized MC exams using their mobile phones[1] to review different ideas about this type of test. I asked students to critically evaluate the information they found on the Internet and reflect on their experience of designing MC questions; each group wrote down what they found informative, and we had a short discussion. I then carried on a whole-class discussion of MC tests and their sociopolitical and educational implications.

In my efforts to facilitate the discussion, I consciously tried to let my students control their peer dialogues to learn from their experiences and interactions with each other. I permitted the entire process to run its course to enable me to determine potential obstacles to students establishing non-dominating and dialogic peer interactions. This process enabled norms and expectations to be established for discussions in future EUPs so that the overall quality of whole-class discussions would be improved.

I asked the class to reflect on their overall experiences of group work, both in our class and in the past. The students agreed that today's group work was productive. Most students enthusiastically shared their experiences and made suggestions to improve the quality of group work. Almost all the students had something to say about group work since they had had some previous experiences of group learning. Based on past experience, the most common drawbacks were as follows:

- Unstructured assignments: Some teachers assign group projects without proper scaffolding and without clearly stating their objectives.
- An overly large group: If it contains more than four people, it is counterproductive.
- A member of the group attempts to dominate: the self-assigned leader issue.
- Some members take credit for the outcome without doing their fair share.

To solve this issue, teachers need to assign each member to a specific task. Students' motivation and participation level were quite high. This discussion validated students' opinions and experiences as they contributed to the learning process in the class (Nystrand, 1997). They articulated negative experiences they had in the past and came up with suggestions for improving the quality of group work in the future. My students were taking ownership of the collective learning process. Most students suggested assigning a specific task to each student in the group to prevent potential problems. This feedback revealed that students had not yet experienced collaborative learning. Group work remained an instrument of learning subject matter. Given this instrumental perspective, the students' suggestion to assign each student to a specific task is reasonable; however, such division of labor is nothing more than working individually in a group, as noted by Horn (2014) and Pietsch (2009). I realized that for the next EUPs, I needed to hold several whole-class discussions to explicitly scaffold students with elements of collaborative learning.

Collaboration and dialogue as conceived by the theory of critical education has certain philosophical roots and ethical implications that drastically differ from group work in neoliberal pedagogy. For example, critical theory ideologically promotes collaboration and solidarity, whereas neoliberal ideology supports competition and individuality. I would have to elucidate this in such a way that my students would clearly understand the difference between these two pedagogies.

Zone of Proximal Development (ZPD)

Although my students indicated some negative aspects of their previous group work experience, they also agreed that learning through collaboration is more enjoyable than studying individually. For instance, Thor said, "My friends may know something, and I don't, and I can learn from them....You know, they may get something from me as well." Students' responses were in accord with Vygotsky's (1978) zone of proximal development (ZPD), which specifies that through collaborative learning students can learn better than learning individually. However, we need to create a classroom ambiance where the students have equal power and where they feel comfortable to exercise their power. If some students hesitate or do not know how to question their peers' ideas, regardless of the norm of equal power for all in the class, other students may possibly dominate in the group or impose their ideas on each other. ZPD outlines a good structure for students to learn from and with each other, but it may potentially create or reproduce power relations among students. I needed to reconsider these points for future EUPs.

Furthermore, to be able to implement ZPD effectively, I needed to identify possible obstacles to students' collaboration. One goal of critical pedagogy (CP) and critical mathematics education (CME) is to turn a classroom into a community of collaborative learners, which requires students to socialize, respect, and work with each other regardless of any cultural, religious, and ethnic differences they might have. However, eight years of teaching in U.S. have made me realize how difficult this might be, as students socialize, in and out of the classroom, with peers from their own culture. For example, based on my lunchtime observations, Asian students sit at the same table while African American students hang out together. In American schools, unfortunately, ethnic pride is mistaken for multiculturalism. This point could be one of many obstacles to establishing a culture of collaboration and dialogue in a diverse setting.

Whole-Class Discussion
I did not want to raise this point for the students yet. Instead, I tried to ascertain if students would bring it up themselves. I asked students whether it would be better if they set their own groups. They responded:

Hannah: Hmm....It is not always productive because we would pick our close friends and be easily off-task and end up chitchatting.

Me: We should be able to work with each other in this class regardless of our ethnicity, gender, race, religion, or any other differences we might have. Our differences would only enrich our life in the class. What is more...what unites us as human beings is much stronger than what seems to separate us.... What do you think?

I observed the students' eyes to capture their immediate reaction to what I said. There was a momentary silence; it seemed obvious that what I just said was new to them. However, they did not appear to disagree. I then asked the class to share their opinion about dialogue and conversation; their responses were interesting. The class agreed with comments made by some that "dialogue" is something that takes place onstage or in a movie; they viewed it as "scripted," as something that does not leave any space for "free talk." However, a few students added that conversation allows one to talk freely; this observation provoked collective agreement among the students. I did not address their comprehension of the term "dialogue," but this incident made me realize that I should continually emphasize the concept of dialogue, collaboration, and inquiry in each EUP. As Groundwater-Smith, Brennan, McFadden, and Mitchell (2003) pointed out, creating a dialogic classroom is a subversive activity—a culture war; therefore it cannot be achieved overnight.

The second topic in the whole-class discussion was whether MC assessment is fair and objective. Students individually shared what they thought, after which a broader classroom discussion began. Nick spoke first:

> I just read an article talking about education in Finland....It says they rarely give multiple-choice tests in school....They focus more on conceptual understanding...and they have the best score on international exams....Maybe we don't need to take so many tests....The article says that American high schools spend about two months of the school year testing.

A moment of silence followed Nick's talk. Having knowingly attempted to avoid IRE and IRF[2] moves, I reluctantly added one comment to Nick's words:

> A comparison between different countries and education systems may open our horizons to understand educational matters better....As you said, there may be no correlation between a successful education and frequent multiple-choice testing....Anyone else wants to contribute?

Ceylan articulated her opinion:

> I don't think it [MC] is a fair assessment because one may know nothing about a question and the other student may have answered almost correctly and made a small mistake just before getting the final answer....Just an example, in our group, we picked a mistake at the last stage that takes third power of negative 3 and writes 8 instead of negative 8....A multiple-choice test degrades these two students....It is not fair at all.

THE STANDARDIZATION MOVEMENT IN EDUCATION

Ceylan formed her conclusion based on her group work experience. The point she raised aligns with similar arguments made by Kohn (2000) and Sacks (2009), according to whom MC assessment applies a binary logic that does not distinguish the answer of a student who has no clue from the response of one who only slightly missed the right answer. At this point, Eric joined in the discussion:

> Let's see...I've never been to a math class like this...you know, like...in math classes, teachers give you equations, and you solve it....We have taken so many tests, but I never thought it was objective or fair, things like that....Ok, we say it was easy or difficult, but that is all....I have just read an article and learned that multiple-choice is cost-effective and practical; that is why we take it so often.

It was interesting that I had already planned to bring up Eric's point, but it came up within the flow of student discussion. As I was preparing a reasoned response in my head, Jennifer objected to Eric's point: "Yeah...but...it may be cost-effective, but that doesn't make multiple-choice good for us....You know, fast food is also cheap and cost-effective, but they are unhealthy." Multiple students' voices were then heard agreeing: "Yay...sounds right." Tibu added another dimension to the discussion:

> Ok, I agree with you [Jennifer]....You may not go to McDonalds, but how would you avoid these [standardized] tests?...Last semester, some of us opted out of EOC [end-of-course tests], but you know the curriculum coordinator lady got us together and told us that if we don't take the test, our school would get less federal funding.

This was an exciting moment for me as a teacher! Eric noticed that this was a non-traditional lesson. He shared an idea from his web search as an explanation for the rationale for MC tests. Jennifer came up with a perfect analogy that contested Eric's point and unpacked the tension between the needs of the system and needs of students. Tibu's comment exemplified how mechanisms of colonizing life operate in the classroom. The discussion seemed to have encouraged my students to build on each other's ideas, aligning with Freire's (2000) suggestion that a dialogic approach and a horizontal student-teacher relationship encourage students to express their ideas freely. This creates the possibility for a dialogic discourse that allows humanizing education to take place in the class. In this sense, my deliberate move to create a non-authoritative and an egalitarian classroom ambiance succeeded: it led to more active and joyful student involvement. This classroom exchange was different from those I had had in the past, and led me to conclude at this point that a communicative space was gradually emerging in the class. I addressed Tibu's comment and carried on:

> In my opinion, we may not be able to change things overnight; it may take a long time. And we have to take and pass these tests in short run. However, this doesn't mean that we should embrace these standardized tests uncritically and unreflectively....As students and teachers—citizens—we have the right to express our dissatisfaction and concerns about standardized assessment....And to be able to change something, we have to have a good understanding of the matter....Who else wants to add something to this topic?

Another moment of silence followed my comment. I sensed that this silence was a sign of the processing that occurs when students are introduced to new or unfamiliar ideas, concepts, or approaches. It is often suggested that moments of entering a new frontier of knowledge trigger a process of inner debate at the individual level (Rogers & Freiberg, 1994; Tsankova & Dobrynina, 2005). Tibu added another comment, returning the class to the central group discussion:

> Mr. Bülent, you are trying to make us think critically....Would not you get in trouble for this? You know, it is like these tests are made by the government, and if you criticize multiple-choice, just like you criticize government, you know.

His tone seemed to be half-serious and half-joking. I did not know how to respond to it at that moment. Upon reflection, however, I was surprised that a student could clearly sense that there is a hidden curriculum. His comment and my reflection on it implied that engaging in an educational practice that fosters critical literacy is not welcomed by official thought.

Tibu's comment and my reflection relate to the notion that market-driven changes have turned schools into places where free thought and critical thinking are not welcome: this is more obvious when students recognize the limitations of non-critical approaches to teaching and learning. Once I gathered my thoughts a moment later, I responded to that question: "We all, as citizens still have a right to express our opinion by the Constitution of the USA, and if making you think critically will put me in trouble, I will take that risk." Directly following my response, Molly addressed Tibu's comment: "He is supposed to make us critically think, why would he get in trouble for that?"

This was a significant moment. Both students are referring to *critical thinking*: while Molly spoke of critical thinking in the sense of functional literacy, Tibu's use of critical thinking generated an element of critical literacy that enables one to think beyond official thought and to talk back to authority.

Darryl raised his hand to join the discussion. He faced his peers when he was talking. At that moment, I was standing behind his table in the left-hand corner and writing down my observations. I realized that this kind of peer communication and whole-class discussion could only occur in a dialogic classroom. I initially decided to address the scope of critical thinking within both functional and critical literacy, but I momentarily changed my mind. Instead, I took note to save this point of discussion for projects to come. I did not want to digress too much to preserve the integrity of the "free dialogue" in which the students were genuinely participating. Darryl stated:

> I was thinking....These tests, like SAT or, you know, EOC, government hires some companies to do these tests, and government pays them for this....It doesn't make any sense to me. Our teachers can make and grade these tests as part of their job....Why involve these companies? And if our teachers graded these tests, government would not have to pay them extra....They already get paid anyway. This seems to me better cost efficiency than paying some big testing companies.

Until Darryl's statement, the students seemed to collectively agree on standardized tests being cost-efficient, but inadequate for measuring students' knowledge. Darryl directly challenged this point, arguing that letting teachers create and score the tests would completely eliminate the cost itself. Darryl's comment emerged as a segue to the wrap-up of the discussion. Many students nodded after he spoke. I deemed it a good time to move on to students' making their journal entries.

Having handed out journals, I posted the following prompt questions for reflective journal entries:

- Provide one negative aspect of your group work experiences in this project and other classes in the past. How do you think we can improve the quality of our group work for projects to come?
- Based on your experience in the project, is multiple-choice (MC) a fair and reliable assessment tool? Why or why not?
- If it is not, why do you think that MC is widely used to measure students' academic success?
- Anything else you may want to address with regards to the project.[3]

Students' Journals

Students' journal entries mostly mirrored the whole-class discussion, but individual students took advantage of the exercise to articulate their thoughts in greater depth. For example, Darryl's comments summed up the whole class's reflection about the project: "Everything in our group work yesterday was great except the problem was a little easy." The simplicity of the task was a point I had noticed in my observation. The immediate lesson I learned as the teacher was to consider more overtly the level of mathematical challenge; a task that was too easy or too difficult would most likely be counter-effective in terms of students' motivation.

However, the approach taken in the project helped some students clarify their misconceptions about some of the mathematical concepts. For example, Cincere wrote, "When we were talking about mistakes and wrong answers, I learned how to figure out shortest distance between a point and a line.... I mean really this time with understanding." Mason wrote, "I always thought that negative sign in front of a fraction would affect top [numerator] and bottom [denominator] numbers....In our group discussion, I got it that negative sign affects top or bottom, not both." These reflections illustrate that in addition to connecting mathematics learning to a larger sociopolitical system, the project stimulated students to improve their mathematical skills and knowledge. The comments also show that the lesson successfully achieved one goal of CME— to ensure that students are engaged in learning mathematics and are successful in a traditional sense (Ernest, 2002c; Gutstein, 2006).

Students' reflections were significant, enabling me develop an understanding of the students' overall perception of the role of group work as compared to more traditional pedagogies that they were accustomed to. Reflections provided me with a starting point toward jointly working on turning our classroom into a community of collaborative learners. Based on their comments, almost all students agreed that domination was a major problem in small-group work. For example, Tibu wrote:

> Group works usually sucks because there is a person or two that assign themselves as group leaders...you know, big shots...and they often control entire process and that sometimes intimidate other people....I believe a solution to this problem would be giving everybody a role based on their strengths and weaknesses, so they can complete their task.

Many students approached the same issue from different perspectives. For instance, Serano voiced another problem: "When we do group work, usually one person works and the others mess around. If each person's role is assigned

THE STANDARDIZATION MOVEMENT IN EDUCATION 19

by the teacher, it may be solved." Bin echoed the same approach: "[If] each person is assigned to a specific task, [it] will solve it." Their comments were informative for me to understand students' current perception suggesting the teacher's domination, since our goal was to create non-dominating and dialogic peer relationships oriented toward establishing a community of collaborative learning.

Many students suggested the teacher's direct intervention as the solution for the problem of domination in group work. These comments seemed to indicate that they prefer to be told what to do, implying that students do not consider themselves proficient in handling potential problems arising in group work. Within traditional pedagogy—which operates as a more anti-dialogic and transmissive teaching (Freire, 2000)—this demand is not unusual. As a result, as part of practicing CME during the next project, I needed to create more opportunities to challenge this expectation and work on developing students' skills and values oriented toward egalitarian and collaborative learning, where students can self-organize, coordinate their projects, and solve potential conflicts that may arise in the process.

Although Lexus brought up the same problem, the solution she suggested was slightly different from those of her classmates: "Domination is always a problem in group projects. I think if everyone in the group is given the same power, this may eliminate that negative aspect." Lexus implies an egalitarian peer relationship is necessary for a truly collaborative learning process. In that relationship, equal power distribution should probably come from the teacher. Bill also acknowledged the domination problem; however, he implied that peers could resolve conflicts resulting from domination without a teacher's direct intervention:

> Sometimes a group member, a know-it-all, dominates the rest and people don't like this. But I believe that this could be solved through a democratic process in the group without outside interference, because usually most people will agree with the most logical and pleasing idea.

Cincere suggested that potential problems in group work should not be reason for choosing to work individually: "When the group is working by themselves and not together and you have two stubborn students in disagreement about who's right. We can eliminate this by settling the disagreement instead of working in singles."

The whole-class discussion and students' journals indicate that students challenged the myth that "standardized assessment is good." As my students developed their arguments, they referred to tests they had taken in the past, their group work experiences, and online searches as they articulated their arguments and conclusions.

Ceylan argued that MC tests may fail at accurately assessing students' knowledge; however, other manners of assessment may also be problematic. She wrote:

> I don't think MC is fair because there is a possibility that a student could get a correct answer purely through guessing and another student could study hard and make a small mistake and end up with wrong answer.... But I think sometimes teachers are biased as well....One of my teachers in middle school, she always favored some students....MC eliminates this.

Ceylan's point is that a student may achieve a correct answer in a standardized test by guessing, whereas a wrong answer may be caused by a minor mistake. However, her comment also reveals skepticism about the impartiality of some teachers.

Bin reflected on the test he took recently to question the reliability of standardized assessment:

> MC is not objective or fair; for example, I just took an SAT test, which is quite important for college application; there was a third-degree equation to solve, I did not know how to solve it? So I just plugged all choices and got the right answer. I got the credit without knowing the subject matter. So how could SAT or other MCs be a reliable assessment tool to define one's future? There should be more options to measure our knowledge.

Bin argued that test results should not be the only tool to define students' academic success and future. On the other hand, Bill approached the test fairness issue from a different perspective: "Based on my experience, MC is fair but not reliable. It is fair because everybody gets the same test, but there is no way to know that the correct answer is chosen by chance or knowingly." Tibu's argument was more of an individual account: "I don't like MC because it is made to expose your mistakes." Cincere framed his argument with a question by writing, "It is so obvious that MC is not fair, but how come people keep saying it works?" He took an important step toward linking life in classroom to a larger sociopolitical system, indicating the contradiction between his experience and the neoliberal argument that claims the standardized assessment is the best way to improve quality of public education.

While most students challenged the idea of standardized assessment without linking it to a broader socioeconomic picture, some students clearly

THE STANDARDIZATION MOVEMENT IN EDUCATION   21

mentioned the political and economic aspect of standardized assessment. For example, Jennifer wrote:

> I think our work should be assessed by free-response problems when we can show what we know and our process....MC is used because it is cheap and quick for the government, but I think our government should do whatever is best for our learning, and they should not use MC just because it is cheap.

This quotation reveals the tension between the needs of students as human beings and citizens and the needs of the system (money and power). Jennifer stated that a free-response exam would be a better way to assess students' knowledge. However, grading a free-response exam is much more expensive than an MC test, as free-response exams usually require certified individuals for grading, whereas MC exams can easily be graded by a scanner. From Jennifer's point of view, a preference for assessment type should be based on whether it allows students to demonstrate what they know rather than whether it is cost-effective. This comment reveals that students recognize that the system imposes its rationale on students and colonizes their life-world in the classroom without their consent. As Apple (2000) and Giroux (2012) point out, in an education system guided by market imperatives, the needs of students are not prioritized.

Jacob was also critical of MC testing. He said that MC does not correspond to the problems that exist in one's life, but he offered a reason for government's preference:

> MC puts you in a situation that you will never face in real life....When problems happen in real life, no one writes five choices for you, and you know one of them is correct....I think testing companies make a lot of money out of these MC tests....It is not a reliable assessment for students, but an efficient way for them [the system] to make money.

Mikael's approach to the fairness of MC was unique. Demonstrating a nice sense of humor, he wrote:

> MC doesn't measure what we know...it focuses on what you do not know....In our group we made four wrong answers and only one correct one. It gives you four opportunities to fail:) how nice! Therefore MC tries to select golden chickens, but it is not fair since everybody wants and deserves to be the golden chicken.

Mikael's response reveals the students' understanding of standardized assessment as an apparatus to produce winners and losers. He thus illustrates the contradiction between the stated argument of neoliberal education to bridge the achievement gap and what students identify as consequences of standardized assessment for themselves (Apple, 2005; Hursh, 2007a).

Having reflected on his group work experience in this project and an online search, Nick made the connection between standardized assessment and its political economy. Like Jacob, he suggested that assessment methods should resonate with challenges in real life:

> When we wrote the question in our group, I realized that MC is designed to trap you, rather than help you....MC does not require deep understanding. For example, sometimes you can plug all choices and get the right answer. In MC, all they care about is the right answer, but life doesn't work that way. There isn't one right answer in life; there are many, sometimes none....In our group, we read an article that was saying that testing and publishing companies promote MC as they make money out of that. I have never heard of that before—very interesting. That explains why MC is still around after all.

Nick based his critique on arguments that link life in the classroom to a larger socioeconomic and sociopolitical structure. Darryl put forward a similar perspective. From his point of view, the government's preference for MC tests is not for academic but rather for corporate reasons. He articulated the contradiction between corporate interests and the needs of students:

> MC doesn't reflect your real skills and tries to mess you up on purpose.... Large companies make MC, and they make lots of money from this....Fair or unfair, they will do everything to keep MC tests like SAT around. But what is good for companies is not good for us [students] and this is a big mistake.

All my students showed their dissatisfaction with MC tests for various reasons. Some referred to their life in the classroom while others made connections between the classroom and a larger political and economic life outside it.

Concluding Discussion of EUP 1

Overall, EUP 1 was a successful entry into the research aimed at studying the potential and limitations of CME in a high school mathematics classroom.

While improving their content knowledge, the students related their learning to a larger society and developed a bottom-up response to neoliberal educational changes. The project helped the students and me to engage in a structural analysis of neoliberal educational policies and implementations. In doing do, we developed and exercised critical mathematical literacy.

In this project, my students and I analyzed standardized tests as part of reflective learning practice. Having created a small opening within the mandated curriculum, we applied a critical approach to understand aspects of standardized tests that are taken for granted and rarely topics of discussion in colonized public spheres. It was evident that through this project that we, as a class, took a big step toward turning our classroom into a dialogic and democratic public sphere.

During class discussion and group work, students listened to each other, shared their experiences and thoughts, and built on each other's ideas. Throughout the project, most students actively contributed in the classroom instead of being passive receivers. As much as I favored the horizontal student-teacher relationships as defined by Freire (2000), my students enjoyed dialogic peer interactions and a more equal structure of participation.

However, I also noted that the skills, values, and attitudes students needed to become active and critical citizens require continual praxis. Based on this first project, some adjustments needed to be introduced for the next phase. For example, students indicated certain individuals' dominance in group work as a potential obstacle to a collaborative learning process. Therefore, in the next project, I needed to work on egalitarian peer interactions to move the classroom interactions toward a more dialogic orientation. Then students themselves could learn about, conduct, and organize a collaborative learning process without requiring the teacher's authoritative intervention. At the beginning, I was both nervous and excited. After I observed the highly positive attitude from my students, I became more excited and hopeful—and less nervous—for the projects to come.

Neoliberal Hegemony in Education

This project helped me to deepen my understanding of the standardization movement in education both in form and content. I reflected on my classroom experience in light of the educational literature on corporate agendas in education. Educational theorists such as Abdi and Carr (2013) and Apple (2014), among many others, consider market-driven changes in education as part of the neoliberal project. A brief historical background of neoliberalism would help us understand who does what, how, and why.

Liberalism was a political ideology mobilized for individual rights against the monarchs in the eighteenth century. Therefore, liberalism, in its historical context, was a progressive step toward emancipatory approaches. The term "liberalism" still evokes a set of concepts—including freedom, democracy, and tolerance—that can be related to a variety of domains, such as culture, economics, and politics, and is generally regarded as a "leftist" world view in the United States and other Western countries (Hedges, 2009, 2010). For this reason, one may mistake neoliberalism as a progressive ideology or at least a non-conservative approach. In the U.S., as Ross and Gibson (2007) point out, "It is important to understand that political conservatives and liberals (in mainstream U.S. politics) both support economic (neo)liberalism" (p. 2). In this broader sense, neoliberalism might be considered as neither a leftist nor a progressive world view.

The philosophical roots of neoliberalism date back to the Walter Lippmann colloquium, which took place over five days in 1938 in Paris. In this colloquium, ideas, strategies, and ideological and political issues were debated with the goal to restore capitalism, which had run into crisis during the Great Depression of the 1930s (Dardot & Laval, 2009). As Hursh (2007b) points out, "Neoliberal economic policies arose as a corporate and political response to the previous Keynesian economic accommodation" (p. 19). The Keynesian approach suggested a welfare state and social net as a corrective to capitalism and to prevent future economic crises. Since the 1980s, through neoliberal polices and implementations, capitalist ruling classes have been trying to undo all these gains of working classes day by day. As such, neoliberalism can be viewed as a historical stage of capitalism.

The neoliberal vision implies a certain political and ideological stance to redefine education in market terms. Therefore, as Giroux (2014) argues, these changes in education cannot be understood without insight into the tenets of neoliberal ideology that devalues "the social, critical agency and informed thinking as part of its attempt to consolidate class power in the hands of a largely white financial and corporate elite" (para. 17). As we attempt to make sense of neoliberal changes in education, we continuously ask ourselves, Whose interests are served by these changes? Then we can arrive at better conclusions.

Neoliberalism targets public schools as part of its ideological-political strategy. Recent market-driven changes in education are not natural events that happen by themselves: rather, these changes are controlled and well-orchestrated events. The strategic goal of neoliberal changes in education is to restore the ideological-political and financial power of the ruling class by turning education into a for-profit business and producing obedient, passive consumer-citizens (the future labor force). While turning education into a for-profit business provides upward flow of money, it is easier to manufacture consent from passive consumer-citizens. We can understand each domain of market-driven educational changes and implementations better in light of the basic tenets of neoliberalism.

Standardized Curriculum and Assessment

The purpose of the standardization movement is not simply to determine the scope and sequence of curricular content for each course, but is part of the bigger plan to reshape public education. From the top down, the standardization movement imposes a one-size-fits-all approach on content and pedagogy. Public school teachers in U.S. have to deal with consequences of standardization in curriculum and assessment. Teachers are relegated to cover a certain amount of preset content in a designated time interval through a certain set of teaching methods referred as "best practices."

This movement imposes a narrow, "teacher-proof" curriculum, aiming to replace education by training (Boyles, 1998). It considers teachers as technicians who deliver pre-packaged knowledge to students, rather than as cultural workers and public intellectuals who interact with students through a dialogic approach (Giroux, 1988; Sacks, 2009). The idea of turning teachers into trainers is based on a market-driven perspective that suggests that schools should be run like businesses. In this view, teaching is something that can be performed by almost anybody who has undergraduate degree; one does not need any pedagogic formation or any intellectual quality. Actually, in some states in the U.S., having an undergraduate degree is indeed the only requirement for getting a teaching position in public schools. The implication is that teaching is just another minimum-wage job, which makes it as easy to hire and fire teachers as it now is to replace hamburger flippers in McDonalds. This business mentality may work to effectively increase profit and decrease cost. But by no means does it apply to education that is supposed to help students become loving, lovable, and caring individuals,[4] as well as active and critical citizens.

A one-size-fits-all approach dictates not only what to teach, but also how to teach. According to this view, a "best practice" of teaching and learning can be scientifically figured out and packaged. For example, in public schools in the state of Washington, teachers have to use scripted words such as "learning targets" and "success criteria" to communicate objectives of daily lessons. Principals visit classrooms with a rubric in their hands to ascertain that teachers put these specific expressions on the board. For teachers who communicate learning objectives in different ways and do not follow these formalities, there are disciplinary consequences.

The neoliberal argument notwithstanding, teaching and learning is a highly context-dependent activity. The context may drastically differ from district to district, school to school; it may be different even within the same school. Teachers would agree that there exists no single "best practice" that fits everywhere and every time. For many years, I have taught the same course in two different periods within the same day. There were times when one class would differ totally from another; what worked in one class caused problems in the other one. Thus, as Hyslop-Margison and Naseem (2007) point out, the market-driven standardization approach denies the fact that students with a diverse set of skills and knowledge have wills of their own; their behavior is usually spontaneous and unpredictable. Furthermore, the subject matter also defines necessary teaching context. For example, I cannot teach mathematics in the same way a music teacher would teach jazz or an art teacher would teach drawing.

The standardization movement also eradicates teachers' academic freedom and authority. Teachers are obliged to cover certain parts of curriculum within a designated time frame. This simply keeps teachers from using classroom time flexibly and covertly urges them to focus on test-preparation activities. If learning with understanding is an explicit goal, teaching a math topic may take two or three lessons for one class but six or more for another. In my experience, when covering as many topics as possible becomes a focal point, rote memorization and superficial learning are almost unavoidable outcomes. In contrast to market-driven implementations, a teacher who is supposed to be an authority on pedagogy and content should have academic freedom to modify curriculum and decide flexible use of instructional time to meet the diverse needs of students.

Standardized assessment with its narrow focus created a big market for multinational testing and publishing companies to make money, but it does not help students think critically about the larger world and their place in it. Giroux (2012) indicates that the standardization movement creates "curricula driven by an anti-intellectual obsession with student test scores, while simultaneously turning students into compliant subjects" (para. 7). This outcome is so visible that two educational theorists from opposite ends of the political spectrum, Henry Giroux and Diane Ravitch, agree on the standardization movement. Ravitch (2011)—who once was a fierce supporter of neoliberal educational changes but later became a strong opponent—points out that market-driven changes promote more testing, more memorization, less critical

THE STANDARDIZATION MOVEMENT IN EDUCATION 27

thinking, and more anti-intellectualism. Therefore, the standardization movement cannot be regarded as a reform.

Accountability System

From a neoliberal perspective, it has been suggested that an accountability system would increase the quality of teaching (Chubb & Moe, 1990). According to this proposal, teachers' evaluations would be based on their students' standardized test results; teachers whose students score higher would receive a little more pay than those whose students score lower. Competitive payment would eventually produce better teachers. Behind this proposal lies a neoliberal ideology of mistrust of public school teachers. It makes the unethical assumption that although teachers have the capacity to perform quality teaching, they do not work to their full capacity; competitive salaries would motivate them to work to their full capacity, and thus they would be better teachers.

The proposal also suggests removing (firing) low-scoring teachers, but fails to explain who would replace them. It is already difficult to become a certified teacher, find a position, and receive a continuous contract: many new teachers quit within the first five years, and those who continue teaching are the most resilient ones. Perhaps proponents of corporate reform assume that best teachers are lined up in front of every public school in U.S. waiting to be hired.

In order to justify a system of teacher accountability, neoliberal ideologues impose, from the top down, ideas that deny any the linkage between students' academic success and their socioeconomic background. From a neoliberal perspective, "The problems of educational achievement and attainment are individual, technical, and scientific rather than economic, social, and structural" (Hyslop-Margison & Naseem, 2007, p. 112). Thus all teachers, regardless of their social, economic, and cultural surroundings, should be subjected to the same accountability system. Proponents of market-driven changes in education, like Bill Gates, argue that students' family background is irrelevant for students' academic success. If schools are run like businesses and effective management is in place, they will be successful regardless of anything else (Chubb & Moe, 1990; Schneider, 2016). This is not a research-based argument: it is strictly an ideological-political statement. Educational theorists and researchers disagree on many things; however, they all agree that the biggest predictor of students' academic success is family background. Denial of this fact is simply nonsense.

The accountability system, in the form of Race to the Top (RTT) grants, measures the success of teachers, students, and schools by standardized test scores. Schools receive less government money if test scores are not high enough (Carr & Porfilio, 2015; Schneider, 2016). By embracing a neoliberal perspective, the No Child Left Behind (NCLB) program and its extension RTT neglect the correlation between students' academic success and socioeconomic background. They deliberately ignore any perspective that draws attention to how the socioeconomic structure of a society reproduces inequality, "because [this correlation] invalidates a conservative and corporate-driven ideological agenda based largely on social Darwinian principles and micro-level accountability" (Hyslop-Margison & Naseem, 2007, p. 106).

That is, advocates for the accountability system claim that scientific research can identify the "best practices" of teaching, which can then be formalized, standardized, and carried out in every classroom, regardless of the socioeconomic class or cultural context that informs students' daily lives (Darder, 2012). This accountability system denies the fact that there are noncognitive factors involved in students' learning and success. Consequently, the process victimizes schools in low-income areas, as they score low and continue to receive less government funding.

Another problem with the accountability approach is that it does not recognize any educational goals or results that are qualitative by nature. A test score is a quantified measurement that cannot, by definition, measure all educational outcomes of teachers' daily instructional and pedagogical practices. Indeed, most teaching practices cannot be measured by numbers. Similarly, important abilities developed through education—abstract, creative, and critical thinking; effective practical skills—cannot be adequately measured by standardized assessment, especially by MC exams.

The system of teacher accountability also covertly pits teachers against each other and eradicates genuine teacher collaboration. My professional experience confirms Smyth's (2011) argument that urging teachers to race against each other creates a toxic ambiance in schools. Each year, schools announce a list of teachers, identifying "successful" and "failing" ones based on standardized test results. However, educational success can only be achieved by a genuine collaboration and solidarity, rather than competition. Teacher collaboration necessitates teachers having academic freedom and autonomy. Because the system compels teachers to see their colleagues as obstacles to their success and a threat to their job security, they do not generously collaborate. Why would they collaborate with their "rivals"? "Professional learning communities" in American public schools have long been places where teachers collectively comply with top-down directives.

Teachers are important part of education process. An education aimed at helping young people to be loving, lovable, and caring individuals and critical citizens necessitates teachers who have a strong intellectual-pedagogic formation and a caring personality. To this end, the system should help pre-service and in-service teachers develop these essential qualities so that we can entrust our children to them. Otherwise, the current accountability system, with its external control and management tactics will continue to demoralize and de-professionalize teachers. Demoralized and de-professionalized teachers may perhaps *train* students to pass standardized tests. However, no miserable teacher can *educate* students to grow as caring individuals and critical citizens. That is for sure.

Students as Future Labor

As a political-ideological view, neoliberalism aims to reshape education to meet the needs of the labor market and thus serve the interests of the ruling class. According to NCLB dogma, public schools in the U.S. are failing, and this failure will negatively affect American dominance in the global market. Therefore, free-market reforms need to be applied to save public schools and produce the future labor force (Bakan, 2011; Kovacs & Christie, 2008; Schneider, 2015). In other words, neoliberal ideologues want schools to be places that produce efficient, obedient workers.

This argument aims to appeal public opinion and sounds exciting at first. However, in reality, "the jobs of future workers [current students] are being relentlessly exported overseas to places with lower labor cost"; thus the claim of preparing the future work force is no more than empty rhetoric or ideological manipulation (Smyth, 2011, p. 15). In a constantly changing world, where it is difficult to predict what's going to happen even a few months from now, who possibly knows what jobs will look like in 10 or 20 years? Thus no correlation can be established between students' success on standardized tests and the skills they will need in the future. Groups advocating market-driven changes in education offer no plausible answers.

Corporations do whatever serves their interests in order to increase profits. There is no direct connection between education and the economy such that improvement in education would improve the economy (Spring, 2014). For example, the great financial crisis in 2008 did not occur because of bad education, but because of the failure of big corporations that were run by highly educated managers and CEOs.

Democracy and Citizenship in Market Terms

Apple (2005) explains what recent neoliberal educational changes mean in terms of democracy: "Consumer choice is the guarantor of democracy"; "the ideal of the citizen is that of the purchaser"; and "rather than democracy being a political concept, it is transformed into a wholly economic concept" (Apple, 2005, p. 273). From the neoliberal perspective, education is valued only in terms of its contribution to the economy, which means that anything irrelevant to market growth "may be legitimately cast off from the ship" (Brown, 2015, p. 211). In this view, classes such as philosophy, music, or film can safely be removed from the curriculum, as they do not contribute to the economy. It is no wonder that in public schools in the U.S., science, technology, engineering, and mathematics (STEM) have been promoted, while humanities and the arts have been downplayed.

Considering the market as a self-regulating force and reducing citizens to economic beings results in shrinking public space for democracy (Blacker, 2013; Cornelissen, 2014). Therefore, when neoliberal educational changes attempt to redefine education on an economic basis, they undermine democracy as a political concept and obliterate education for critical citizenship. As Giroux (2012) elaborates: "Students in this view are no longer educated for democratic citizenship. On the contrary, they are now being trained to fulfill the need for human capital" (para. 5). In other words, in the eyes of the power elite, our children are nothing more than the future labor force. It is important to remember that in the U.S., children of the power elite do not attend public schools. They instead go to posh private schools where education is far different than what market-driven changes envision for students in public schools.

As a result of market-driven changes, public schools in the U.S. are moving away from being public spheres where students-as-citizens can negotiate, learn, and experiment with skills, democratic values, and knowledge to be able to be active and critical citizens to keep authorities accountable. Schools have instead become places that are colonized by corporate agendas, where students are seen as objects for economic growth from which the ruling class benefits the most. From a neoliberal perspective, as Hyslop-Margison and Naseem (2007) note, "Students are portrayed as mere objects in history and ideologically inculcated with a consumer-driven world view devoid of imagination, hope, or alternative social visions"; they are expected to adapt themselves to existing conditions rather than "engaging with or democratically transforming their political, economic, and social landscape" (p. 121). From this one can conclude that corporate school reform inspired by neoliberal ideology envisions students as employable future workers who can perform to solve given business problems, but not critically question the socioeconomic and sociopolitical structures that affect their lives.

Education plays a vital role in developing and sustaining a democratic society (Chomsky, 2003; Dewey, 1916). Therefore, the overall objective of public education is to provide students with opportunities of becoming critical citizens who can actively participate in the process of decision making that affects their individual and social lives, and who are competent to hold authorities accountable. Curriculum and pedagogy should be integrated to foster certain democratic skills, attitudes, and values (Giroux, 1989; Hyslop-Margison & Thayer, 2009; Ravitch, 2011; Skovsmose, 2011). However, neither NCLB nor RTT seems to recognize democratic values and the role that education plays in cultivating them.

School Choice and Charter Schools

Neoliberal advocacy groups manipulated and exploited many concepts with which American people have positive associations: *choice* is one of them. Discourse of school choice takes center stage in market-driven changes, which ironically leaves most vulnerable part of society with no choice. In this view, standardized (allegedly objective) test scores inform parents which schools are "best" (i.e., have the highest test scores). Then parents have a choice: they shop around for the best schools and send their children to these successful schools wherever they are located (the cost of transportation belongs to parents).

From this market-based logic, Chubb and Moe (1990) argue, parental choice will trigger competition among schools; hence, good schools will survive and bad ones will close down. As a result, schools, regardless of their surrounding socioeconomic and cultural conditions, will get better across the country, and the gap between advantaged and disadvantaged students will disappear. Every student would receive a high quality education; every student would go on to graduate from college; the U.S. would continue to dominate the global economy (Chubb & Moe, 1990). Neoliberal advocacy groups have appealed to some parents and the wider public in general at a surface level, promising to provide the best education for their children and promoting U.S. dominance in the global market.

In contrast to neoliberal advocacy claims, since the NCLB Act was initiated in 2001, there has been only limited school-based evidence to suggest that public schools have responded positively to market-driven initiatives (P. R. Carr & Porfilio, 2015; Hill, 2009; Hursh, 2007a; Kohn, 2004; Kovacs & Christie, 2008; Ravitch, 2016). In fact, research on the impact of market-driven educational changes indicates that "low-income students are less able to take advantage of school choice programs due to costs of transportation and information" (Lincove, 2009, p. 3)—a condition also reflected in the poor educational achievement of working-class students of color (Darder, 2012).

Similarly, studies undertaken by S. Ball (1993), Orlowski (2012), and Abrams (2016) indicate that market-driven changes in education have benefited privileged groups of society and victimized poor and low-income groups. This outcome clearly contradicts the neoliberal argument, which promises to bridge the gap between poor and well-off students (Apple, 2014; Kohn & Shannon, 2002; Kozol, 2005; Schneider, 2016). There is only limited school-based anecdotal evidence of positive outcomes due to neoliberal changes in education. However, such evidence comes from a couple of charter schools that are well funded by advocacy groups and used as showcases.

Giroux (2014) points out that public education is targeted by neoliberal globalization, just because it is public. From a neoliberal view, education is an individual commodity rather than a public service or a human right, and thus it can be bought and sold like any other product in the marketplace (Coco, 2013). By this premise, schools would be more effective if they were privatized and run like businesses. *Politics, Markets & American Schools* is almost a sacred book of neoliberal advocacy groups. In the book, Chubb and Moe (1990), proponents of neoliberal approaches to education, frame a neoliberal perspective to point out the failure of public schools, putting forth privatization as the only viable alternative to fix education. In their view, democratic governance of American public schools produces bureaucracy, which is the main source of their failure: "Democratic control normally produces ineffective schools" (p. 227). Chubb and Moe propose that schools should be still funded by taxpayers, but run by private companies. Market-driven changes in education have also produced a great deal of bureaucracy as a top-down system. Standardized curriculum and assessment mean that teachers have to deal with mountains of paper work; this takes a considerable time and energy that can be used for working with students, improving instruction, and the like. But neoliberal advocacy groups seem to have no explanation for that.

Currently, democratically elected school boards govern public schools in the U.S. The board decides how and where taxpayers' money is to be spent as social investment. The school board's decision is accountable to public control as board members are elected. What neoliberal advocacy groups want to achieve is to create a channel that will smoothly transfer taxpayers' money to corporations' bank accounts. They propose that boards of trustees—who are not elected but promoted, and therefore free of public control—should run schools; private companies can make a profit out of this business. As a public school teacher who has seventeen years of experience nationally and internationally, I can make a list of problems with American public education; however, not a single one necessitates handing over the control of public schools to private corporations. Corporate takeover of public education has

THE STANDARDIZATION MOVEMENT IN EDUCATION   33

made the problems worse and will continue to do so as long as the market-driven perspective is in effect.

Who Are behind Neoliberal Educational Changes?

It is interesting that both Republican and Democrat parties have supported market-driven changes in education. Major corporations and their side organizations are behind this privatization initiative. In the U.S., for example, the Bill & Melinda Gates Foundation, the Heritage Foundation, the Thomas B. Fordham Foundation, the Manhattan Institute, and the Milton and Rose D. Friedman Foundation are some of major organizations that sponsor private think tanks. These think tanks, in turn, are fierce supporters of the neoliberal movement in education; they initiate various activities in academic, political, and cultural domains to influence public opinion for privatization of public schools.

These neoliberal advocacy groups in the U.S. operate mainly in three domains to legitimize, implement, and institutionalize market-driven educational changes. First, through lobbying, they have already persuaded the federal government to force each state to accept, implement, and institutionalize market-driven changes in public education (Bakan, 2011; Ross & Gibson, 2007; Schneider, 2016). The implementation of common-core state standards (CCSS)[5] can be seen as the most recent example of this process (Schneider, 2015). Second, they have been running campaigns to manage public perception[6] in order to manufacture general consent for neoliberal educational changes (Scott, 2011). Third, they aim to influence educational literature, promoting empirical research and scientism in order to reinforce the hegemony of neoliberal ideology in the academic world (Hyslop-Margison & Naseem, 2007). When principals, teaching coaches, and district officials speak about policies in public schools, they use specific language associated with market-driven discourse.

What about Neoconservatives?

One may wonder where the neoconservatives stand on this matter. Although neoliberals have led and are still leading this educational transformation process, they are not alone. Educational theorist Michael Apple (2000) considers this educational change as a *conservative* restoration process: "The conservative restoration [is] guided by a tense coalition of forces, some of whose aims partly contradict others" (p. 72). Apple argues that conservatives are the second driving force in this coalition: while neoliberals are for weak state power, neoconservatives stand for a strong state; cooperation between these parts might appear unlikely.

However, as Apple points out, the two factions of the ruling class overcame their differences when their interests overlapped. From the perspective of critical theory, this alliance can be understood to be between neoliberals and neoconservatives, two capitalist ruling classes in the U.S. While their political approaches may differ, their class interests are usually in harmony. Privatization of public schools benefits both parties as they belong to the same capitalist class. In that sense, the coalition between neoliberals and neoconservatives is not contradictory, as Apple (2000) explains: "As odd as it may seem, neoliberal and neoconservative policies that are seemingly contradictory may mutually reinforce each other in the long run" (p. 73). For example, the standardization movement created a big market for publishing and testing companies—"edu-business." At the same time, this process enabled conservatives to prevent educational movements such as bilingual education initiatives. Cooperation between the Democratic Party and the Republican Party for market-driven changes in education can be better understood in this context.

Having critically assessed the implications of market-driven changes in education, we arrive at an answer to the question, "Market-driven changes are positive (or negative) for whom, from what point of view, and by what measure?" While none of the neoliberal arguments for educational change has yet been verified by research (Ravitch, 2016), there is evidence that market-driven changes have created a new sector, "edu-business," from which multinational publishing-testing companies and educational consultancy companies genuinely benefit (Abrams, 2016; Boyles, 1998; De Lissovoy, Means, & Saltman, 2015; Hursh, 2007a; Kohn, 2004; Kovacs & Christie, 2008; Schneider, 2015). Neoliberal educational changes have had deleterious consequences for schools in low-income areas at large, but they genuinely benefited big corporations.

Do We Have Any Alternative to Neoliberal Changes?

In broad terms, critical pedagogy (CP) is a radical response to neoliberal educational practices that are oppressive and exploitative. CP defines pedagogy in terms of class, ideology, power, economics, culture, and politics. From this point of departure, CP promotes educational practices that help students and teachers detect root causes of the oppressive and exploitative dynamics of traditional education (Darder, Baltodano, & Torres, 2009). To this end, CP continuously searches for emancipatory and empowering alternatives to traditional education so that teachers, together with their students, can develop critical consciousness and become agencies of change toward a more just, equal, and free society.

CP radically criticizes neoliberal educational changes in American public schools and elsewhere in the world. Recent studies in CP point out that public schools in America have been colonized by neoliberal pedagogy. Through its micromanagement and control strategies, the market-driven perspective in education dictates how to teach, what to teach, and when to teach in the classroom. In doing so, teachers are stripped of their authority, autonomy, and academic freedom. Neoliberal pedagogy turns teachers into trainers who are supposed to deliver "teacher-proof" curriculum. CP, then, concludes that the neoliberal hegemony eliminates possibilities of democratic practices of education.

However, CP also concludes that schools colonized by neoliberal pedagogy need the practices of CP more than ever. Although CP has a strong and established literature, it offers few concrete ideas that would help high school teachers counter neoliberal hegemony. Critical educational theorists such as Michael Apple, Henry Giroux, Peter McLaren, Antonia Darder, and Ira Shor, among many others, have produced valuable works in CP. However, they are mostly theoretical studies and lack classroom-based data. Ironically, critical theory puts a huge emphasis on the dialectical relation between theory and practice. Without this essential connection, however, most of the studies on CP unfortunately become theories about theories that never touch base with life in the classroom.

Critical Mathematics Education

Critical mathematics education (CME) can be seen as a subset of CP. A small but increasing number of scholars in mathematics education are focusing on the sociopolitical and socioeconomic aspects of teaching and learning mathematics (Frankenstein, 1983; Gutstein, 2006; Skovsmose, 2011). These studies are united in positing a radical critique of traditional perspectives. CME aims to foster critical citizenship and catalyze transformative social changes (Frankenstein, 2010; Gutstein, 2006; Skovsmose, 1994; Valero & Zevenbergen, 2004). The works of Skovsmose (1994, 2011) and Skovsmose and Alrø (2004) are especially important in providing a coherent foundation from which to define a practice of CME. There are some classroom-based studies that draw on CME in the high school context (Brantlinger, 2014; Gutstein, 2006). However, currently no classroom-based research exists in the CME literature that frames a practice of mathematics education within the high school context to challenge neoliberal imperatives.

In response, the present book, through rich classroom-based data, reconciles CP's contradictory stances and provides a reconceptualization of CME as a counterhegemonic practice of teaching and learning mathematics in the presence of neoliberal restrictions.

Notes

1 My students and I agreed on the rule: No talking on the phone or texting in the classroom. Cell phones in this project were for educational purposes only.
2 Initiate-respond-evaluate (IRE) and initiate-respond-feedback (IRF) are regarded as nondialogic, transmission-style instruction.
3 The intent of this question is to allow individual students to propose agendas and discussion points they might have.
4 I borrowed these words from Noddings (2003).
5 By 2014, some 49 states had adopted a standardized curriculum; the website http://www.corestandards.org/ provides information on standards for each content area
6 The documentary *Waiting for Superman* is a good example of the efforts to create the impression that public education is failing, and that "bad" teachers (and the unions that protect them) are responsible for the failure.

CHAPTER 3

Class Consciousness and Mathematical Literacy

This chapter explores ways in which mathematics instruction can provide students with opportunities to be engaged in sociopolitical analysis of work-life and class-based society. It examines the ways in which an open-ended word problem can contextualize conflicting interests in a class-based society. The chapter emphasizes that socioeconomic background is a strong identity marker and that class consciousness is essential for one to become an active and critical citizen. It describes the students' response to the sociopolitical situation they find themselves in, where their life-worlds are colonized by neoliberalism, and where undistorted and free discussions about class society are prevented from happening.

The chapter also argues that existing anti-racist movements in the U.S. are domesticated. If an anti-racist movement, academic or social, is to counter neoliberal hegemony, it must be integrated into a broader movement that prioritizes structural inequality and class. I argue that CP should reclaim class as a primary unit of analysis in order to counter neoliberal hegemony.

Planning and Objectives

After EUP 1, we began a new unit of district-mandated curriculum that covered the mathematical concept of *function*, including topics such as domain; range; graphical, numerical, and algebraic analysis; piece-wise functions; and family of functions. In this project, students needed knowledge and skills concerning algebraic, numerical, and graphical analyses of a function. As well, they needed to learn how to calculate intercept point(s) of given functions. We had already covered the mechanical aspects of this content prior to the project. Therefore, students were familiar with these topics as we began EUP 2.

I developed EUP 2 in light of the lessons learned from EUP 1. The main theme of this project was socioeconomic class division and inequalities. Within this theme, my intention was to focus on public interests versus corporate interests. I reviewed word problems in mainstream algebra textbooks to design a word problem. However, I noted that almost every word problem in these textbooks involved themes such as best-buying decisions, maximizing profit, or minimizing cost; my intention was to find a word problem that could be revised. But I could not find any word problem that would be useful for this project. So I designed an word problem of my own. I developed EUP 2 by applying these four criteria:

© KONINKLIJKE BRILL NV, LEIDEN, 2019 | DOI:10.1163/9789004390232_003

- The word problem should not be so open-ended that students lose focus and motivation; however, it should not be too easy, either.
- It should contain rigorous mathematics at conceptual level in connection with our current unit as defined by the district standard to meet procedural necessities.
- Students should be able relate the problem to their life-world.
- The problem should be inquiry-driven so as to inspire students to collectively engage with critical and creative thought and have space for dialogue and negotiation through which they can approach the problem from different perspectives as they work in groups.

The word problem for this project is shown below:

> *Project 2: Part-Time Worker*
>
> Edward is a senior student in High Hill High School and has recently got a part-time job in a local restaurant to support himself and his brother Joel, who is a middle school student. He is going to work there after school and sometimes over the weekends. As he negotiated his weekly salary with his boss, his boss offered him two different calculations for his time:
>
> $S(t) = 16t + 20$
>
> $W(t) = (8/5)\, t^2 + 20$
>
> Edward's total work time cannot be more than 40 hrs as he is a part-time worker. Before Edward made his final decision, he told his boss that he needs a day or two to think about it. Also he asked his boss whether or not he could combine these two offers as a multipart function of time (or piece-wise); the answer was YES....It could be a multipart function. Edward will collaboratively work with his friends in his math class to come up with an offer that best presents his interests out of all possibilities.
>
> *Solution*
>
> The problem illustrates a negotiation between a young worker who is a senior student in a high school and his boss. The boss offers the student a weekly salary, which includes two distinct functions. The problem allowed my students to develop different solutions that favor either the young worker or his boss. The purpose here was to challenge the common perception that mathematics problems always have a single answer (Skovsmose, 1994, 2011). Further, because CME is aimed at fostering critical literacy, my intention was to contextualize mathematical discussions (functional literacy) in a discussion of social class issues to promote critical literacy.

Day 1

To begin EUP 2, I facilitated a whole-class discussion. Having reflected on the previous project and students' feedback, the students and I together decided to negotiate and set some norms for working collaboratively. First, we addressed the students' reflective journals; there were several suggestions by students to improve the quality of group work. For example, some students had proposed that if the teacher assigned each student in a group a specific task, it would solve problems of domination. I briefly explained that if I assigned each student in a group a specific task, this would no longer be qualified as collaborative learning; this is evident in the following transcript:

Me: Our goal is to achieve learning and applying math through collaboration and dialogue....Dialogue is much more than our daily conversation. In our group work, if we try to dominate each other...if we do not listen to each other...if we do not build on each other's ideas and suggestions, our conversation would not evolve into dialogue....Without dialogue, our group work would not turn into a process of collaborative learning.... It would be just an individual's work in a group.

Nick: That sounds right, but what if one tries to run over everybody or not work at all?

Me: That is what we need to figure out together in our class....I have some suggestions, and you should come up with your own as well.

Jennifer: Ok....Honestly, I sometimes step back [in group work], especially when I am not confident about subject matter.

Multiple students: Yep!...That happens to me....Me, too.

Me: Can we, please, talk one at a time....[a moment of silence]. Then it is our responsibility to welcome our friends who, for some reason, are left out....Similarly, if one tries to dominate, politely remind him/her that we are trying to achieve helping each other rather than racing each other.

Lexus: I think we should all agree that everybody in group work should have same power....I don't like when people are bossing me around.

Me: I absolutely agree with Lexus....If anybody wants to add something more...

In this brief discussion, the students agreed to jointly coordinate group work on their own and ask me for suggestions as needed. Further, we decided that if a student was not part of collaborative learning, other group members should gently try to welcome them in. Finally, the students jointly agreed that if somebody tries to direct the group, they would be politely reminded that our goal is non-dominating collaborative learning. The students' comments revealed that they recognized that dominating behavior in group work is not productive as it disrupts the inclusiveness of group work, which aligns with suggestion made by Horn (2014) that true collaborative learning necessitates egalitarian peer relationships.

The discussion turned out to be much more stimulating than I had anticipated. The students' comments indicated that they enjoy the humanizing, dialogic, and non-hierarchical learning process described by Freire (2000). As Freire argued, whereas dialogue allows the development of a humanizing education, anti-dialogue is the root cause of oppression characterized by the "banking" concept of education.

After this brief discussion, I handed out the project paper and had five minutes of silence for students to review and process the word problem. Then I divided them into mixed-gender groups; each group contained students from diverse backgrounds. Circulating among the tables, I took notes about peer interactions in groups. As I was in one corner of the classroom, from the other end, Tibu raised his hand and made a joke, which was loud enough to be heard by everyone in the class: "Mr. Bülent, Ben is trying to dominate our group, and he is very dogmatic too." Multiple students laughed at this comment. This joke generated a warm ambiance and signaled students' awareness of what we were trying to achieve and their enthusiastic participation in this collective learning process. A group of three seemed to be engaged in lively conversation. I noted that Allen was making a T-chart to see the input-output relation; Jennifer was trying to graph the two given functions with a graphing calculator; and Nick was solving the system to find the intercept point. The next transcript shows the development of the group's problem solving through peer dialogue:

> Allen: I think Edward should go by the linear one....It seems he makes more money this way...
>
> Nick: Are you sure? Quadratic one should bring him more money.... You know each time you square, it curves up quickly...
>
> Allen: I tried 2, 3, 4, 5, 6 hours...definitely linear one is a better choice for Edward...

CLASS CONSCIOUSNESS AND MATHEMATICAL LITERACY   41

Jennifer: Actually [showing graphing calculator's screen to her peers], you guys are right....Linear one has a higher output first, but it changes after a point.

Nick: You mean the intercept point? I got 10, is that what the calculator shows?

Jennifer: Let me check....Yes! It is 10....Allen, can you plug 12 in your table and see what happens?

Allen: [he calculated by pencil] Wow! Now quadratic one is way higher.

The excerpt indicates that the students learned from and with each other as they worked in each other's ZPD (Vygotsky, 1978). Peer discussions gradually evolved into dialogue; each student followed a different method in approaching the problem, which seemed to complement one another, and no student was left out.

In one particular group, each student applied a different method to solve the problem, and they jointly combined their answers to develop a piece-wise function that favored Edward. I noted that peer communication in group work was inquiry-based and egalitarian. It appeared that no student was trying to compete with the others in their group. It was also evident that in this incident, peer solidarity was an emerging and self-organizing phenomenon.

All groups seemed to be engaged and making contributions to completing the task. Furthermore, no student was excluded during discussions. I focused my observation on one specific group for a while; the group was already discussing how to frame a multipart function in favor of Edward. I noticed that no group was trying to come up with a multipart function in favor of Edward's boss. It seemed that students had a strong empathy with Edward as they could relate his situation to their own life-world. I came to this conclusion after reflecting on a question I had posed at the beginning of the school year. I asked students if they were working somewhere to make money. Twenty-six out of 28 students had a part-time job either after school or on weekends.

Since we emphasized non-dominant, non-hierarchical peer relations in group work, students seemed mindful and sensitive in their interactions; all students enjoyed that their participation was equally valued. They took ownership of the learning process and generously contributed to their group without trying to be a power figure.

My observation strongly confirms Freire's (2000) point that vertical relationships or hierarchies—either among peers or between the teacher and students—eradicate possibilities of participation and consequently dialogic (humanizing) learning. Once egalitarian peer interactions occur, students feel more confident and can productively socialize with each other and learn with and from each other, as suggested by the Vygotskian concept of ZPD. However, unlike Vygotsky's notion, there is no need for a more competent student to tutor others.

After the problem-solving process was completed, two groups briefly shared their solutions on the board. Noah, Taylor, and Michael's group presentation seemed highly effective in terms of the quality of collaboration. Noah explained, "I did equalize two functions to calculate the edge point for piece-wise function." I intervened and asked if this was their first action, since many students seemed to use a T-chart to make numerical analysis. Taylor answered: "I was trying to do that [T-chart/numerical analysis] first, but Noah said it would take a long time." Nick (who was not in the group) raised his hand and said, "I believe this was a collaboration, like it's an 'us' project, not a 'me' project." Jokingly, he added, "I think Noah attempted to impose his perspective on his group."

This humorous perspective spontaneously created a warm and welcoming atmosphere; students laughed for a while. Taylor said, "It may look like Noah was dominating, but I don't think he meant it." Michael added that they should have politely reminded Noah that each voice was supposed to be heard. Michael also agreed that Noah's act was not intentional. I intervened with a question: "Can we generalize Michael's comment that when authorities or government attempt to impose something, it is our duty as citizens to come forward and voice our concern that we are not ok with the situation?" After a short silence, Noah's comment finalized this conversation: "I had no intention to dominate my group. I just said that a T-chart would take a long time and is not always reliable, and I can still show that a T-chart takes much longer than solving systems of equations." It was evident that Noah was trying to say that his action was driven not by power dynamics, but a mathematical argument.

Egalitarian collaborative learning and dialogic peer interactions were part of the objectives of this project. The conversation revealed that my students were learning and embracing dialogic and non-dominating peer interactions in their group work. Therefore, this outcome indicated that we as a class had made progress toward objectives set for this EUP and for the entire project. As Pine (2009) suggested, developing dialogic and collaborative learning has to be learned, and it requires a systematic approach.

Nicole's group also shared their group work. She placed her group's paper under the document camera and explained how they came up with a graphical presentation of the situation. Jacob went on to explain algebraically and

CLASS CONSCIOUSNESS AND MATHEMATICAL LITERACY   43

numerically why their multipart function favors Edward. At the end of the explanation, Daniel asked the class if they had any questions. Eric raised his hand: "This Edward's scenario does not look like real....I mean, if he works 30 hours a week, he would make a lot of money....No job would pay that much.... What were you guys thinking?" As Jacob attempted to answer the question, he was interrupted by a school-wide announcement by the principal, and we had only a couple of minutes left for that period. I quickly tried to wrap up discussion, "Ok, let's continue tomorrow....We'll come back to this question as soon as the period starts. Please turn in your final work before you leave the class. One paper work for each group....Thank you, and have a nice rest of the day." I instantly noted that this question would be a good start-off point for a whole-class discussion next day.

Day 2

Next day, the lesson began with the question from the day before: Was the word problem compatible with current wages? Because on the previous day Jacob was cut off by the announcement, I invited him to start. He said, "If Edward worked 30 hours a week, his annual salary would be way higher than a part-time job would pay. I myself work in an ice-cream shop and get paid minimum wage." Allen added, "Even many full-time jobs don't pay that much at all." My students were already linking Edward's story to a larger economic system that seemed to frame this project as a meaningful learning experience.

Darryl raised his hand: "It is not only wage....I think it is not realistic that Edward's boss offers him a choice. People usually work under restricted conditions." Selena joined in: "I don't think his boss will accept our piece-wise function that favors Edward. I mean, why would he?....The boss has the power and Edward doesn't. Why would any boss agree to pay more?" Tibu responded, "If the boss wouldn't accept the function that favors Edward, what could his motivation be? Like...is he trying to see how smart Edward is?"

These comments reflect a successful realization of one of CME's fundamental goals—to ensure that students are provided with learning opportunities to develop critical mathematical literacy and are able to ask questions such as "whose interests are served and who benefits?" (Gutstein, 2006, p. 5). I momentarily intended to make a transition from a relation between the boss and the worker in the context of Edward's story to public versus private interests in the context of a corporate-driven world. However, we had time restrictions, and needed to complete the project today. We still had to make entries in our student journals. I shared this concern with my students, and we then moved on to a journal session.

44 CHAPTER 3

Students' Journals

I posted the questions shown below on the board using the projector. As a class, we agreed to spend ten minutes for a discussion of these questions in groups before making journal entries. While some students focused on a single question, others responded to multiple questions. I let my students know that in connection with this project, they could bring up any point that they considered worthwhile—their responses should not be limited by the prompt questions. Because this action research was conducted with, not on, my students, they should also be able to come up with their own questions to address.

- Do you think a student needs basic mathematics skills in order to make informed and educated decisions and to be critical/active citizen?
- Do you think that we live in a class society?
- Do you think Edward's interest clashes with his boss's interest? Why and how?
- Can you give an example of private versus public interest?
- Do you think that the quality of collaboration and dialogue in your group work for this project was better than our previous project?
- In what ways was this project mathematically meaningful for you?

The journals were collected five minutes before the bell rang. I asked the students if they would like our next project to be on the history of mathematics. They enthusiastically asked me several questions. I told them that if we agreed, it would be an open-ended inquiry and each group could decide their own topic.

Within four days, I reviewed all journals. By referring to Edward's employment story, my students clearly articulated that one needs to be literate in mathematics to be become an active and critical citizen. Nicole illustrated the developing perspectives on how critical literacy could allow individuals to prevent sociopolitical manipulation:

> Basic math and even basic school education is necessary, in my opinion, so one is not manipulated by others, but is an active citizen instead.... If Edward did not know graphical and functional analysis, he may have ended up with a package that favors his boss. In our group, we came up with two piece-wise functions....One was in favor of Edward, and the other was good for his boss.

While Selena agreed with others on the necessity of basic mathematics skills for becoming an active citizen, she argued that advanced math skills are not necessary for one to become a critical citizen. She stated, "In order to be an active citizen, [and] also to survive, you need basic math skills; however, not everybody has to know advanced complicated math." This point was also a challenge for CME: Where do we draw the line between basic and advanced mathematics in the context of critical citizenship?

Jennifer added another layer to the discussion: "I don't think one needs only basic mathematics skills not to be manipulated or be active citizen; I think one should be able to speak for oneself." Her comment implies that functional literacy itself may lead to adaptation to a given socioeconomic condition. However, one needs critical literacy to demand change for the better, as well as courage to stand up for one's rights. She questions whether there is a relation between knowing something and the courage to act upon that knowledge.

Darryl agreed on the necessity of mathematics skills; he looked at the issue from a wider perspective and provided examples to back up his point:

> I do believe that one needs basic math skills in order to not be manipulated by others and to be active citizens. Math is everywhere in our lives. This includes our economic exchanges. For example, when purchasing insurance, one must know about all the different parts and fees as well as what they will cost and cover to make an informed decision that is best for you....Large companies are smart and know how to take advantage of people to gain profit....Having basic math skills helps one not to fall in these traps. One must also have these skills to make informed decisions when voting on taxes and budgets.

Darryl's comment reveals an emerging critical mathematical literacy: He appeared to be aware that corporations have no self-imposed limitations when it comes to making a profit. One needs critical (mathematics) literacy to detect and challenge these situations, which aligns with the definitions of critical mathematical literacy made by Frankenstein (2005).

Lexus's comment focused on Edward's story. She wrote:

> People like Edward need to have basic skills. If we take this scenario like real, if Edward did not know his math, his boss could have easily manipulated him into a piece-wise function that benefits the boss. If you have basic understanding of math, you will be able to pick the one that benefits you.

My students' journal entries indicated that they believed mathematics knowledge and skills to be necessary for survival and for being active and critical citizens. They agreed that they live in a class society by providing evidence from their life-world. However, in the U.S., the matter of class is a sensitive subject to discuss in public. According to Chomsky (2003), there is a top-down imposed perception that Americans live in a classless society. Thus, I did not anticipate that my students would be able to smoothly connect Edward's story to the larger socioeconomic system. I was expecting that at least a couple of students would repeat clichés such as "This is a land of freedom and equality... we are all same," etc. However, as I evaluated the results of this project along with the previous one, the dialogic, collective, and inquiry-based learning process in the class was becoming a more self-directed process. The students' reflections indicated that inquiry-based dialogue could create a communicative sphere where even a top-down perception—such as that of the U.S. as a classless society—could be challenged.

In her response, Jennifer identified asymmetrical power relations in society to explain why we live in a class society: "No matter how free and classless we think we may be [in our country], some people have more power than others. There are bosses and workers; the rich and the poor; and the government and the people." From her point of view, a classless society would require equal power distribution. Sidney's reflection was complementary to Jennifer's, linking Edward's story to the significance of critical literacy for ordinary people to make informed decisions:

> Yes, we live in a class society where high-class people have lots of money and education and power to manipulate for their benefit; therefore, people like Edward should have math skills and other skills as well not to be deceived or exploited.

Sidney implied that in a class society, those who are in power position may exploit ordinary individuals' ignorance; therefore, citizens need critical mathematical literacy to protect themselves as much as possible from these possible deceptions and exploitations. Her journal entry does not imply that being literate in mathematics would prevent deception and exploitation of all kind. She refers to Edward's story to support her point. Molly goes beyond Edward's story and directly refers to her everyday experience in the school to address to the question:

> Obviously we live in a class society....Even in our school, one can see that there are students who live in families where [they] don't always get a full meal every day or warm clothes to wear. One can also see the students who always have whatever they need whenever they want.

Molly was the only one who pointed to the socioeconomic background of children at our school to exemplify the class society. Even though most

CLASS CONSCIOUSNESS AND MATHEMATICAL LITERACY

students in my school come from a poor socioeconomic background, some are from higher-income families; my long experience in the school confirms Molly's argument. The students' comments revealed that critical mathematical literacy was emerging as they examined their lives and Edward's story in relationship to sociopolitical and socioeconomic contexts, in the same vein as Gutstein (2006, p. 5) distinguishes critical literacy from functional literacy.

During this project, we did not have any direct discussion about clashing interests in a given society. However, the students' comments indicated that the word problem had two meaningful solutions. While one favors Edward, the other would favor his boss. They noticed that the solution that benefited the boss was not good for Edward. For example, Leonardo made several connections to indicate that we live in a class society, including a recent campaign[1] in Seattle, Washington, to increase the minimum wage:

> Yes, we live in a class society. The rich people's interests and regular people's interests are quite opposite. For example, while workers in the Seattle area demand higher minimum wage, $15, which is currently $9, the rich oppose this, as a wage increase means their profit will decrease. The same thing for Edward....If we make a piece-wise function to take side with Edward, his boss will get less profit. On the other hand, if we favor the boss, Edward will get underpaid. However, the point is that Edward doesn't have any power, while the employer has money and power over others.

Leonardo's comment revealed that Edward's story enabled him to develop different approaches to the problem. This outcome resonates with Skovsmose's (1994) suggestion that word problems should be designed in ways that provide a landscape of investigation for students to develop multiple answers as opposed to a single correct answer. Leonardo, who had a part-time job to meet his school expenditures, noticed the basic fact that employers have asymmetrical power over employees; he implied that Edward's story in this project does not reflect reality as far as power relations go.

One of the prompt questions listed earlier invited students to think about public interests and corporate interests in order to make a transition from the interests of employer and employee to corporate interests. We did not have much time to discuss this point during whole-class discussion. However, some students established a coherent logic to make the transition. For example, Ceylan addressed the question in the context of Seattle's minimum-wage campaign:

> An example of class interest in our society is the upper class arguing that minimum wage should not be increased because it will only cause inflation. Obviously, it would be in their [corporate] best interest to keep the working class's salary as low as possible, so they don't lose revenue in paying employees more....Public interest is in favor of the overall good of society as whole. Private interest is in favor of one individual's personal gain like owners of big companies....I am not sure how salaries exactly are determined, but it is probably determined by the bosses who calculate the minimum amount of money they can give to their workers that will pay them enough to sustain themselves while maximizing the boss's personal profit.

Looking at socioeconomic structures through the lens of the working class, Ceylan's reflection revealed her understanding of critical mathematical literacy as she identified oppressive and exploitative aspects of labor relationships (Freire, 2013; Giroux, 1983; Gutstein, 2006).

The students linked this EUP to an ongoing public discussion over wage increase; it was one of the tangible successes of practicing CME here. In fact, I was going to do one EUP on the wage-increase campaign in Seattle, but the school principal (head teacher) did not approve it: I was told that this was a controversial issue and may cause problems. Therefore, I did not raise this topic in class. Nevertheless, some students made the connection in a different context by linking Edward's story to the larger socioeconomic and political structure.

Creating a dialogic classroom culture through inquiry-based and collective learning was the ongoing goal. We attempted to make group work a more democratic and collaborative learning process. My students reflected on the quality of group work as it was experienced in the project. Nicole acknowledged that we had improved: "The group work in this project worked a little better than the previous one. We worked very well together and did same amount of work without overpowering one another." Nick provided a more detailed account:

> All members of my group were actively participating, asking questions to each other, listening to each other's questions, pulling their own weight, putting out good work and ideas....As working on the project proved that we count on each other to do their parts, but also we respect everyone and try not to be dogmatic and suppressive. Everyone wants to be able to work in an equal, non-dominant environment, and I think we achieved just that in this project....First time I felt, honestly, that I got why we need piece-wise functions and how intercept of two graphs could be useful.

CLASS CONSCIOUSNESS AND MATHEMATICAL LITERACY 49

His reflection indicates that in this project, dialogic peer relations and collaborative learning of mathematics have improved; this notion seems to be consistent with that of Pietsch (2009), who argues that establishing dialogic collaboration requires systematic effort and time.

Kohn (1992) indicated that collaborative learning creates a sense of belonging and community; Jennifer's reflection resonates with this notion:

> I felt like a part of my group in this project was way better than last time....I spoke my ideas on how to find edge point for the piece-wise function without thinking, What if I was wrong and they would make fun of me?

Jennifer felt that she would not be judged or embarrassed if she made mistakes. It is evident that her group achieved respect and tolerance, which are vital for dialogue and collaborative learning. Darryl also indicated that their group work was a success:

> Hannah, Nick, and I communicated very well to make all calculations and algebra to figure out the edge point [intercept point] to graph the system and write a multipart function for Edward....We greatly contributed to each other's thoughts as well as the final work itself.

In my observation, I noticed that Edward's story triggered inquiry-driven peer collaborations. As I had anticipated, the problem called for multiple approaches to possible solutions; the transitions from mathematical argumentation to students' life-world to a larger socioeconomic system were dynamic.

Race, Class, and Critical Pedagogy

When I designed EUPs, I used my professional preference as the classroom teacher to choose the theme of each EUP. As an educator who embraces CP in his teaching, I was initially going to include racism as the theme of EUP 2. But, having reviewed the relevant literature and reflected on my conversation with students and overall classroom experiences, I changed my mind. Instead, I chose class-based society as the main theme of the project, which, for several reasons, is more a inclusive topic than race.

Anti-racist movements in the U.S. made historically progressive gains and created significant cracks in hegemony of racist discourse in 1960s and 1970s. When we review the rhetoric of leading figures of these struggles such as Martin Luther King and Malcolm X, we see that the social, political, and economic structures of the system were central to their criticism and political actions: they were radicals. King publicly showed his solidarity with all oppressed people, not just with black community; he even expressed his empathy with the Vietnamese people who were tortured in Saigon prisons during the Vietnam War. In other words, the anti-racist movement at that time was part of a larger straggle toward freedom, equality, and justice for all.

However, since the early 1980s, the anti-racist movement lost its momentum and larger emancipatory vision. In academic studies in the 1990s, especially, race became a primary unit of analysis with an unclear political vision. Simultaneously, in the social sciences and humanities, *class* as an identity marker seemed to be forgotten (Orlowski, 2012). Studies under the rubric of race theory produced concepts such as "multicultural education," which has different meanings for different people. For example, in my experience, what the average high school student understands from multicultural education is more or less ethnic pride or chauvinism.

These movements promote equal access to educational credentials for black students. In other words, black students should have equal opportunities with white students. Which white students? Do all white students have the same opportunity? As we all know, the answer is No! White working-class students have significantly less educational opportunities and resources than students from well-off families. Then how are we to equalize black students' opportunities with white ones?

As a public educator and researcher, I contend that the power elite in the U.S. accommodated the initial demands of the anti-racist movement but then turned them into politically ineffective activities. With some exceptions, the ruling elite usually managed to domesticate anti-racist academic and political movements. Unfortunately, some black scholars and activist groups, as well as some left-wing intellectuals—especially those who embraced postmodern identity politics—provided the ruling class with a great deal of (unintentional) help in this process. Neoliberal ideology used race- and ethnicity-based identity politics to divide working-class people.

From this point of departure, I question how current anti-racist individual activism and social movements are framed; what they bring to and take away from larger straggles against neoliberal hegemony. I agree with the political stance that claims that race and class issues are inseparable in the U.S. Therefore, a counterhegemonic social movement or pedagogy should prioritize class and focus on the ongoing struggle between labor and capital. Unfortunately, anti-racist movements that problematize race-based discrimination against the poor working-class black community but fail to recognize neoliberal capitalism as the root cause of misery of the working class in general,

CLASS CONSCIOUSNESS AND MATHEMATICAL LITERACY

are at the service of ruling class. No matter how radical these movements may sound, they ultimately reproduce the existing oppressive and exploitative system because they lack the political-ideological vision to counter neoliberal capitalism.

The ways in which racism is addressed in schools (and in society) usually generates a perception that reduces racism to individual matters. A perception that considers all white people as potentially racist arrives at absurd conclusions, such as white teachers cannot teach black students, or vice versa. The perception is equally limited that blindly assumes all white people are well-off, ignoring structural inequality and exploitation that targets white working class students, and thus alienates them in school. For example, I had two students: one black, the other white. Both worked at McDonald's after school, and coincidently their parents were employed in a local Wal-Mart. The neoliberal system, which enforces the upward flow of wealth, makes victims of both these students, not just the black one.

A counterhegemonic teaching practice must decode the ruling class's strategy that aims to pit these two students against each another. This pedagogy must cultivate critical literacy so that black and white students can come together along with other oppressed groups in society (schools) to fight for their common interests against the common enemy. In other words, we need a whole new language and understanding of racism that distinguishes between problems of black and brown people in the back streets of America and the problems of a black Hollywood star who complains that no black actors or actresses were nominated for an academy award. While the Hollywood star's race-related problem could be solved within the system, solutions to problems of working-class black people require radical and revolutionary changes.

During professional development days and educational workshops initiated by the state or school districts, analysis of success of students, schools, and teachers usually includes students' racial and ethnic background; however, students' socioeconomic background is never taken into account. No official statistics in education relates students' academic success to their family's socioeconomic background. In the colonized life-world of Americans, class is an unspeakable topic, a "no-fly" zone. Subversive teaching practices should allow classrooms to be places where students *can* speak about class matters in social and political context. That was my point of departure when I chose class-based society as the main theme of EUP 2. In the word problem, I portrayed Edward as a working-class student. I did not mention his ethnic or race background, nor did any student ask for that information. All students had a part-time job and each of them could possibly be Edward.

Concluding Discussion of EUP 2

In EUP 2, all the students worked on establishing dialogic peer interactions. Based on students' reflections, inclusiveness in group work was achieved. They had previously said that when someone acts as a self-assigned leader, others might be intimidated or discouraged. Once non-dominating peer interactions have matured, students seemed to become more careful not to exclude anyone.

The students appeared to develop a democratic and egalitarian sensitivity—as witnessed in their reflections—and at the same time improved their critical mathematics literacy as they connected their life-world in the context of Edward's story to the larger socioeconomic system. Although I provided scaffolding for the process, I aimed to be an equal research partner with my students. The dialogic classroom environment gradually created a self-regulating framework, which led to small openings within which students could question concepts such as the class society and differences between private, public, and corporate interests. This project met the four objectives listed earlier and thus was another step toward creating a dialogic mathematics classroom and successfully practicing CME.

Note

1 Since 2014, there has been a civil movement for increasing minimum wage from $9 to $15 in Seattle. This movement triggered a larger public debate in the Seattle area and spread to other parts of the U.S. Further information can be obtained at http://www.occupydemocrats.com/seattle-mayor-says-behind-living-wage/

CHAPTER 4

Collaborative Versus Competitive Learning

In this chapter, I explore ways in which history of mathematics can be integrated into critical mathematics education. I argue that students should be given opportunities to learn how to learn collaboratively and reflect on their experience. As students develop more agentic voices, they are able to articulate that collaboration has been shown to be more beneficial for the human family than competition. Neoliberal pedagogy, on the other hand, promotes a culture of competition. Therefore, fostering the collaborative learning of mathematics is a way of practicing critical mathematics education and countering neoliberal pedagogy.

Planning and Objectives

The day before we began EUP 3, I taught a brief lesson on operations with zero and infinity. I wrote mathematical expressions on the board (as shown below) and then facilitated a whole-class discussion to encourage students to develop some insight into these topics. My intention was to evoke their curiosity about the history of mathematics by briefly introducing the historical background of these concepts.

$$\text{a)} \frac{3}{0} = ? \quad \text{b)} \frac{0}{3} = ? \quad \text{c)} \, 2\infty \pm \infty = ? \quad \text{d)} \frac{3}{\infty} = ? \quad \text{e)} \frac{\infty}{3} = ? \quad \text{f)} \, \infty \times \infty = ?$$

This brief lesson incorporated an instrumental aspect. The next unit contained the concept of asymptotes, which required working with the notions of zero and infinity. Students usually have difficulty comprehending topics that require a deeper conceptual understanding than the mere procedural knowledge needed for solving skill-drill-type repetitive examples. I facilitated a brief class discussion to negotiate each key question. For example, students' first reaction to expression (d) indicated that they assumed infinity was a known number. The discussion helped students to clarify some confusions and misconceptions. However, from the stance of critical mathematics education (CME), to help students comprehend such concepts in more depth, it would be necessary to explain the historical context in which each concept emerged.

This project is the continuation of EUP 1 and EUP 2, which aimed at creating small openings in our mathematics classroom for students to develop values, attitudes, skills, and knowledge to become critical citizens. Similar to the

© KONINKLIJKE BRILL NV, LEIDEN, 2019 | DOI:10.1163/9789004390232_004

previous two projects, we intended to achieve these goals through inquiry-based collaborative learning and dialogic classroom interactions. D'Ambrosio's (2010) approach to mathematics education was my main inspiration for this project. The objective of this unit was to provide students with the opportunity to realize that mathematics is a collective production of humankind and recognize that collaboration historically is more beneficial and less destructive than competition.

D'Ambrosio (2010) conceptualizes critical mathematics education as "ethnomathematics." He argued that mathematics education could contribute to world peace and provide students with the ability to survive with dignity:

> As a mathematician and mathematics educator myself, I accept as a priority, the pursuit of civilization with dignity for all, in which inequity, arrogance and bigotry have no place. This means, to achieve a world in peace and to reject violence. (p. 53)

Inspired by D'Ambrosio's (2010) point, I considered the history of mathematics the main topic of this project. I asked the students if they would like to inquire into the history of mathematics. As students requested some examples, I told them that they could research, for instance, the contribution of non-European cultures or the historical emergence of certain mathematical concepts such as pi, zero, or infinity. The students were excited to be able to select a research topic for their group.

Day 1

I posted the details for this project via electronic projector and then explained the scope of EUP 3. I handed out the paperwork as a hard copy and let the students review it for five minutes. The project included a problem-posing aspect; it did not directly involve any mathematical problem in a conventional sense. Students were supposed to collectively determine the topic of their study and its extent; they were able to use Internet links provided in the package to do the research. Therefore, in comparison to the previous projects, this one required higher coordination skills, collaborative decision making, and dialogic peer interaction in order to achieve inquiry-based learning.

In the previous two projects, I set the groups. This time, responding to the students' emerging collective initiative, I allowed them to select their own groups. Each group was provided with two laptops with Internet access and a printer. As I circulated among groups, I observed that students were becoming collaborative, deciding what to do without excluding anybody. Trying to dominate one's peers and expecting the teacher, as an external authority, to

COLLABORATIVE VERSUS COMPETITIVE LEARNING

assign each student to a specific task were no longer notable issues—they all participated and contributed equally to the work. However, as I reviewed the lesson package at the end of EUP 3, I noticed that the project was too open-ended. I did not specify where to end research on pi, for example. Each group had to decide where to start and how to conclude, which could be frustrating and confusing for students.

Nevertheless, the students were more self-organized than they were with previous projects. They were excited about collaborative learning for EUP 3 and enjoyed taking ownership of their learning. I observed that they handled the entire process very well: with each project, the students were becoming more active as participants and increasingly being the agents of their own learning. Each group had a brief discussion and agreed on a topic from the packet. They prepared posters out of their research by including everyone in the group. The quality of questions they asked each other and ways in which they responded made me think that there is a dialectical relation between dialogic peer interactions and collaborative learning: one would not exist without the other. My students improved their skills of active participation and taking initiative. Observing this made me feel great and gave me hope for a better future as a public school teacher. My students were interacting with each other differently, changing their ways of learning and doing mathematics.

It was evident that students' ability for collaborative learning and peer dialogue had improved since EUP 1. I realized that inquiry-based egalitarian collaboration and non-dominating peer interactions had been transforming the class into a community. The reciprocal was also true: As the class became more of a community, egalitarian collaboration and dialogic relationships among students were more apparent.

An incident occurred during this project that was significant in terms of highlighting some limitations of CME within a neoliberal educational system. While students were working on the project, the school principal visited our class as a part of my teacher evaluation. He talked with the students in each group and took notes on his laptop. At our school, teachers receive their evaluations through an Internet-based system. Four days later, I received an email with a link to review my evaluation. In the comments section, the principal left the following note:

> I talked with most students in the class. They were extremely engaged with their projects. They answered my questions with excitement. I reviewed the lesson plan, but I could not see any standards. I wonder what common-core state standards were addressed in this lesson.

Even though an instrumental part of the project focuses on the content, I was not able to relate this lesson to any standard in the paperwork for the project. In fact, the new national standards, common-core state standards (CCSS) do not include any information about history of mathematics. There was also a section in the evaluation form for me to respond to the principal's comment. I recorded the same comment in my journal:

> The standard (CCSS) for high school mathematics, unfortunately, does not cover history of math. Therefore, I was unable to directly refer to a specific standard. However, the lesson indirectly relates to learning objectives in the standards. For example, students need conceptual understanding for usage of zero and infinity to fully comprehend concepts such as undetermined situations, undefined fractions, and asymptotes. In this sense, the lesson covered multiple standards. Also the National Council of Teaching Mathematics (NCTM) strongly advises integrating history of math into math classes to empower students. Moreover, in all developed countries, the math curriculum includes history of math. In my professional opinion, our curriculum should cover the historical dimension of math. I believe that it is our responsibility to provide our students with the opportunity of experiencing cultural and social side of math beyond repetitive equations, isolated facts, and theorems even though current standards do not suggest doing so.

This incident forcefully reminded me of the intricate daily dialogue between neoliberalism, standardization, and innovative teaching and learning; it confirmed the critical analysis of neoliberal education by numerous scholars (Apple, 2000; Hursh, 2007b; Kohn, 2000; McNeil, 2009). Commenting on the restrictive aspect of neoliberal educational changes, Apple (2000) noted that some states in the U.S. "not only have specified the content that teachers are to teach but also have regulated the only appropriate methods of teaching. Not following these specified 'appropriate' methods puts the teacher at risk of administrative sanction" (p. 70). This episode reminded me that I needed to be more cautious in balancing the needs of the standardized curriculum as I practiced CME.

I kept circulating around the class and observed groups learning about different symbols for numbers, having discussions about pi, zero, infinity, and the connection between mathematics and technology. Toward the end of the class, Malik asked: "My brother is taking calculus, and I looked at his textbook. Every chapter cites all these individuals who developed calculus. But our algebra textbook doesn't cite anybody. Does algebra come from an unknown history?

COLLABORATIVE VERSUS COMPETITIVE LEARNING 57

What do you think, Mr. Bülent?" I did not expect such a question, and I was not sure if Malik had his own answer and was just checking on me. I responded, "Why don't we research this question tonight when you go home? We may be able to see different opinions and approaches to Malik's question. And we can add this question to our class discussion tomorrow."

At the end of the class, students suggested posting their work on the wall instead of doing presentations. Their rationale was that when we did presentations, only one or two groups could present their work due to time constraints. On the other hand, if they posted their work on the wall, it could remain there for much longer, and other students in different periods could read them as well. It was a strong idea that involved students in the decision-making process, which resonates with the spirit of participatory action research. All the groups hung their posters around the classroom walls and talked about each other's projects.

Day 2

Students were ten minutes late because of a fire drill from the previous period. Having briefly reviewed the material from the previous day's class, I facilitated a whole-class discussion. Because students selected different research topics, I opened the discussion with a broad question:

Me: Mathematics is a joint production, and humankind and we should appreciate it; however, math has not always been used in the past to make our life better. Perhaps we need to look at math beyond our classroom walls and see how it functioned in past. If you need to say something about your research yesterday, what would like to share with your classmates?

Nick: It was interesting that math people in ancient times visited other cultures to exchange ideas. They did math collaboratively just like we do in this class.

Multiple students laughed for a while. There was a sense of community spirit spontaneously generating a warm and inviting ambiance as they began thinking and talking about the history of mathematics:

Lexus: Yes! Greek mathematicians went to Egypt to study math, for example.

Jacob: It was amazing to see how different cultures used different symbols for numbers and did addition and multiplication with two-digit numbers...way different than we do today.

At this point, I tried to redirect discussion onto specific questions.

Me: Do you think mathematics has been more of a help or harm for people?

Bin: We would not have any technology without math, and you know technology is something good for humanity. But I am not sure it is always beneficial for people.

Ceylan: We read an article talking about nuclear wars, and atomic bombs would be impossible without math. I guess math can be used to harm people, but it can be used for good causes. For example, math was used to make weapons and bombs to kill people. But the energy of a bomb could be used to heat up households during wintertime.

Me: I agree with Ceylan. It depends on who uses math with what purposes. A corporation may use math to increase their profit while communities can use math to optimize their needs based on their resources for public good....We learned that math is an outcome of collaboration, but we all know that people may not always work together. What do you think keeps people from collaborating instead of making them fight or compete?

J-Paz: If people have different [conflicting] interests, they would probably not collaborate. That makes sense, but I never understand why people fight with each other when their interests are the same. If they help each other, it would be in everyone's favor.

These comments revealed that the students considered collaborative learning as more productive and less distractive than competitive learning, a discovery that resonates with a comprehensive analysis by Kohn (1992). Kohn argued that whatever can be achieved through competition can also be achieved by collaboration. Unlike competition, collaboration is not destructive. The discussion continued:

Me: What about our classroom; do we have any rational reasons for not collaborating?

Darryl: It is different....Our collaboration is for learning, and, as Jen said, we all want to learn and we learn better when we study with each other.

COLLABORATIVE VERSUS COMPETITIVE LEARNING

Sidney: I personally feel good about myself, I mean, whenever I learn something new from my friends or they learn from me. It makes me feel I am part of this class.

Me: We learned in this project that mathematics is a joint production of the human family. What would you say if a friend of yours came up to you and told you that he/she was proud to belong to the human family?

Tom: I would tell him to go to the nurse.

[Multiple students laughing]

Ceylan: It sounds like a cheesy word to say, but we actually belong to a human family. My friend may be a Filipino, she may be a Korean or Mexican, but we are all human beings.

Andrew: Yes...but, I feel different, like, you know...I belong to my own culture. I am German-Irish. If you say human family....You know, you also include Chinese or Russian or something else, and I am not part of their culture.

Ethnicity and race are delicate topics in the U.S. I carefully attempted to come up with an analogy to convey my point:

Me: Well...if you think of your ethnic background as a flower in a big flower garden, your flower looks nice as it is situated next to others; it enriches the entire garden. Without the big garden, your flower would not be the same. As you are proud of the big garden, you are also proud of your own flower while appreciating existence of others.

These discussion questions took up almost the whole period. However, I felt as if we were jumping from one topic to other. I had planned to address the question that Malik had asked yesterday, but instead I included his question in the prompt for journal entries as we ran out of time. On that day, we had a 15-minute-shorter period as the office scheduled a student assembly at the end of the day. I posted the prompt questions on the board and gave students 10 minutes to exchange ideas with each other. I then handed out their journals. They quietly made their entries for the rest of the period.

Prompt Questions for Student Journal Entries
– Would you consider mathematics as a joint achievement of all cultures of the world or specific nations/cultures? Explain your rationale.

- In your opinion, can differences be barriers for people preventing people from collaborating to produce common good? Provide examples to back up your answer.
- Do you think that it would be a good idea to make history of mathematics an integral part of our mathematics curriculum? Why or why not?
- While all calculus textbooks cite their contributors, algebra textbooks do not mention the historic background of algebra. Can you think of any reason behind this?

I anticipated that some students would think mathematics was a Western production as opposed to a product of many nations and cultures. But all students stated that mathematics emerged from a diverse human history. For example, Taylor perceived mathematics as a universal value:

> Before this project I had always thought that math was something unto itself; through this project, I realized that it is a collective achievement of humanity. Math is something that brings people together like music....I also learned the concepts of zero, infinity, and pi and their philosophical stuff.

Lam drew on an argument that opposes the idea of a cultural superiority. Given that all the world's cultures contributed to mathematics, this could not have been done by one single nation or culture. He also asserted that dialogue is needed to work with others productively, and people's differences do not have to be barriers for dialogue:

> This project made me think that no culture is superior to another culture....Throughout history, different cultures contributed to the development of math. We did research [in our group] and learned that some mathematicians in ancient times visited different places, civilizations, and exchanged ideas to improve math....One thing I got from this is that differences should not keep us from having dialogue with others.

Some students made clear connections between mathematics, culture, and politics. For example, Ceylan shared her ideas about the underlying reasons for cultural superiority of Western countries. She linked advanced technology to mathematics and their emergence in the Western world. However, Ceylan also indicated that without other cultures' contributions, advanced mathematics or technology would be impossible:

> Yes, all cultures contributed to math, but advanced math resulted in more advanced technology, which generally correlates to feelings of cultural

superiority like developed Western countries. But these Western countries developed advanced math on top of what other cultures had already done, like algebra came out in Babylon. Western countries are politically strong and have huge military power, but I don't think that makes them culturally superior to other nations.

Ceylan's comment was consistent with that of D'Ambrosio (1999), who suggests that integrating the history of mathematics into the curriculum may motivate students who come from historically marginalized and underserved communities. Jacob used the history of mathematics as an example to support his conclusion that collaboration is beneficial: "It made me think that without collaboration, math would not be a math as it is today....We agreed in our group that, by any measure, I think, collaboration produces more common good than competition." It appeared that more students began to see value in collaboration, and the classroom became a better community of collaborative learning.

Reflecting on previous EUPs, I facilitated whole-class discussions in EUP 1 and EUP 2 to probe common obstacles to collaboration. Students provided valuable feedback and suggestions to overcome those obstacles. Collaboration had always been voluntary. I gave students the option to work individually if they wished to do so. Interestingly, however, all students preferred to work with their peers. The class as a whole considered collaborative practices to be useful for their learning. In their journal entries, students reflected on the dynamics of competition and collaboration. Darryl argued that our common roots are stronger than our differences:

> Some people think that differences are obstacles for collaboration to achieve common values. I don't agree with that. We are all human beings [and] should understand that we come from the same beginning. We should respect each other to work together to change the world for the better for everyone.

In Darryl's view, there is no reason to consider people's differences as obstacles to collaboration; such differences can be overcome. Nicole stressed the irrationality of the notion of cultural superiority. Feeling superior to others eradicates possibilities for collaboration and dialogue:

> Cultural, religious, and other kinds of differences do not have to be obstacles to work collaboratively, but sometimes it can be as well. When people from a certain cultural/ethnic background see themselves superior to others, differences become barriers for collaboration. On the other

hand, interacting with different people can actually teach you something. Learning to appreciate others' ideas and values can also make it a value of your own.

From these comments and others like them, I concluded that having a discussion on collaboration versus competition helped the students and me gain a deeper insight into potential obstacles to collaborative work. It also helped make their collaborative activity more conscious and deliberate during small-group work. Nicole implied that the notion of cultural superiority feeds arrogance, whereas one needs to be humble to appreciate others' ideas and values.

Jennifer-Paz put forward a perspective similar. She seemed to consider our collaborative learning efforts as an outcome of a conscious joint decision. She implied that we have to overcome being dogmatic and dominant:

> I would like to think cultural, religious, or ethnic differences should not get in the way of working collaboratively, but the reality is [that] it becomes an obstacle sometimes. The sad but inevitable truth is as long as people remain opinionated, differences may keep people from working collaboratively. We do collaborative learning in this class, because we agreed on non-dominating communication, and we are trying not to be dogmatic.

Some students, however, appeared to be skeptical about humankind's capacity to achieve peace. For instance, Cindy wrote:

> I am not so sure if differences are barriers for people to work collaboratively to produce common good. Yes, we have universal values that unite us like love, art, music and math, but how can we explain all these wars and bad things that took place in history?

She seemed to realize that collaboration is positive and beneficial for everyone. At the same time, she recognized the violence and injustices of the past, thereby avoiding a romantic optimism. Her line of thought can be considered to contain a healthy level of skepticism rather than pessimism.

On the first day of this project, one student asked why algebra textbooks do not cite their founders or historic background, while every calculus textbook credits mathematicians who contributed to the development of calculus. Students' responses to this question were beyond my anticipation. When I reviewed their journals, I realized this question would be a great point of departure in order to discuss cultural superiority, world peace, collaboration, and dialogue. Malik, who had raised the question earlier, felt strongly about it:

> Algebra textbook is not credited because it is not from the Western hemisphere....Anything that comes from non-Western countries is not important in our [U.S.] culture....Even when we use things like yoga, we degenerate, commercialize, and strip it of its authenticity and roots.

Hannah agreed with Malik. She wrote, "In our country, things are valued if they come from Western cultures/countries. Otherwise, it is either ignored or belittled. That is why the founder of algebra is not cited in our algebra textbooks."

Reviewing the students' journals, I also noticed that students whose cultural roots are non-Western strongly believe that their cultural existence is marginalized and not valued. Most of them were born in the U.S., but that does not change their feelings. This was a very delicate situation about which I exercised caution: A sense of being marginalized in a classroom (or in a society) may be the result of chauvinism or other discrimination. A society in which people have no dialogue would soon become atomized individual consumers. Therefore, through critical mathematics education and critical pedagogy, I strive to help my students develop universal values to deconstruct and reconstruct their thoughts, skills, and values to become critical citizens.

As a mathematics teacher, I can confirm that algebra textbooks in the U.S. do not cite any historical figures or backgrounds. For example, it would be extremely unusual for an algebra textbook to cite the historical background of the quadratic equation. However, calculus textbooks credit the founders of calculus whenever appropriate. Calculus was invented in seventeenth century in Europe, whereas algebra was created by Islamic scholars in the near East in the tenth and eleventh centuries.[1] The founders of algebra are known figures, and they could be credited if authors and publishers chose to do so.

Students explained the choice to not credit those scholars by referring to the ongoing tension between two worlds: the East and the West. Nicole said, "I think Western dominance is the reason why the founder of algebra is not credited in our algebra textbooks." I realized that integrating ethnomathematics into daily lessons, as D'Ambrosio (2010) proposed, could effectively challenge the Eurocentric vision of mathematics. Helping students recognize that mathematics is a joint production of the human family aligns with the ideas of Frankenstein (1990) and Powell and Frankenstein (1997), who argued that integrating the history of mathematics into the curriculum would empower students.

64   CHAPTER 4

Despite the fact that EUP 3 was only a two-day project, the students' reflections showed that it seemed to expand the their horizons and make a positive difference in their life-worlds. Molly shared her overall impression of the project:

> Learning HOM [history of mathematics] will promote conceptual understanding. I had never understood the concept of zero and infinity until this project; as we had discussions, I now better understand why a number cannot be divided by a zero. Also I had never thought of math as being potentially used for bad things such as nuclear warfare. This was an interesting point of our discussion; how do we keep people from using math for bad?

Molly posed an important question: What can we do to keep mathematics from being used to harm people? Similarly, Ceylan implied that the project gave her a new perspective on the formatting power of mathematics in the modern world. She wrote, "I never thought about the ways mathematics could be used for positive and harmful purposes." These comments resonated with Skovsmose's (2011) point that mathematics may lead to both "wonders and horrors." He provided examples of both: "It is very difficult to think of any medical research without mathematics playing an integral part....Military enterprises can not be carried out without mathematics" (p. 68). He suggested that ways in which mathematics affects our world should be subject to critical reflection and ethical evaluation.

Concluding Discussion of EUP 3

As demonstrated by their comments and journals, as well as the whole-class discussion, the students embraced the imperative to be active and reflective participants in the learning process. My role in this project was no more than a facilitator; this is in line with Rogers's (1995) suggestion that the teacher's role is crucial to establish a dialogic classroom. As the students' participation was encouraged and valued, they took ownership of the learning process. For example, instead to having presentations—with some groups not having time to present—they proposed hanging posters on the wall to extend the discussion and include every group. Their highly reasonable suggestion was a turning point, showing that our classroom was becoming both a community of collaborative learners and a dialogic classroom.

COLLABORATIVE VERSUS COMPETITIVE LEARNING 65

The whole-class discussions and journals indicated that, in the students' view, collaboration produces more common good for humanity than competition does, as evidenced by the historical development of mathematics. However, students also developed this view through experiences in the projects, gradually noticing that they learned and felt much better doing peer collaboration. Their reflective understanding of dialogue, respect, and solidarity helped them construct this view, as action and reflection are mutually reinforcing.

This project helped students consider mathematics from a historical-political perspective, as opposed to simply being an educational subject, and helped them revise some of their ideas. In the case of the history of algebra, for instance, they problematized the notion of cultural superiority. I did not know that some of my students felt strongly about this dominance. However, their reflections did not reveal any sign of aggressive ethnic pride. I was expecting some students to make chauvinistic arguments; fortunately, that did not happen. Instead, they talked about universal values, indicated world peace as a common interest of all people, and opposed the idea of cultural superiority and arrogance. In that sense, we partly achieved our objectives.

Nevertheless, from a larger perspective, I realized that the objectives of EUP 3 were too broad, which resulted in superficial learning. For example, idea of peace and history of mathematics in the context of citizenship was a very broad topic. Two days were clearly not enough to cover all of the objectives. But I had no chance to spend any additional time on this project, as we had to cover a certain number of topics within each month as mandated by the standardized district curriculum. Despite good intentions, looking at EUP 3 from this perspective showed this project to be, to some degree, a fast-food style of education. I realized that the history of mathematics should be incorporated into every unit throughout the year.

Moreover, the principal's classroom visit reminded me of the ever-present tension between one's life-world and the system, as described by Kemmis et al. (2014). While my students and I were trying to decolonize our life-world, I neglected fulfilling the system's requirement by not including codes for learning standards. The principal's observation of our classroom (and the project) could lead to a very lively discussion between the classroom teacher and the principal beyond technical formalities such as referring to the standards. However, as part of its control and management effort, the system was more concerned with the codes for learning standards—regardless of whether the lesson was actually educationally beneficial for students.

In order to create small openings in our life-world (our classroom), I need to keep my job. Teaching a lesson that was not linked to district standards did not cause any serious consequences for me on that occasion—I received a verbal warning that I should not teach anything not covered by the curriculum. However, under different circumstances, it could have led to serious consequences, resulting in a poor teacher evaluation or a letter of direction. I realized that I needed to be more careful and creative to meet system requirements while reorganizing to democratize our life-world.

Note

1 Further information about this topic can be found at http://www.storyofmathematics.com/islamic.html, as well as many other websites and books that deal with the history of mathematics.

CHAPTER 5

Mathematical Inequality and Socioeconomic Inequality

This chapter describes the fourth action research project (EUP 4) and illustrates how linear systems of inequality can be contextualized to address socio-economic inequality. I argue that neoliberal ideology considers sociopolitical and socioeconomic matters as individual problems and rejects viewing them at the societal level. The chapter explores the ways in which a math teacher can make small openings for students to have undistorted and free discussions about community volunteer service, social inequality, poverty, and solidarity.

Planning and Objectives

The mathematical content of EUP 4 concerned linear inequalities. Prior to the project, we studied mathematical inequalities and practiced mechanical aspects of the subject matter, so that students were able to solve given systems of inequality by graphing or by algebraic approaches. However, my intention was to develop a project relating systems of inequalities in mathematics to inequality as a socioeconomic concept. Therefore the class critically reviewed concepts such as selfishness, generosity, failure or success, charity, solidarity in context of class society.

I planned to use a word problem that I had developed and used in my classes many times. In the problem, students initiate a campaign for raising money with certain restrictions to serve their community. But I was not completely satisfied with the context of the problem or its linkage with critical pedagogy. As I was simultaneously working on my professional development, I reread an article by Freitas (2008). One of her word problem examples was a system of inequality, and it inspired me to revise my word problem in a way that would incorporate social issues such as homelessness, charity, generosity, poverty, community service, solidarity, and social responsibility, in order to counter neoliberal pedagogy.

I framed this project as another story about Edward, whose name was featured in EUP 2. To begin the lesson, I handed out the following word problem.

Poverty in the U.S. is a delicate issue; blaming the victim is the prevailing ideology. Whenever a poverty or social inequality issue comes up, average people

© KONINKLIJKE BRILL NV, LEIDEN, 2019 | DOI:10.1163/9789004390232_005

Project 4: Spirit of community and solidarity

Edward, along with his friends David and Jane, volunteered for a community organization called *Solidarity Here and Now* (SHN), which is a non-profit organization, run by the local community and aims to support those who are in need. SHN owns a building, which is used as temporary housing in winter sessions. Edward, along with his friends David and Jane, are in charge of this housing project. This is a two-story building with foldable (accordion) walls: It is 6,000 square feet that could be used for different purposes. In summertime, it is used for community activities, and during the winter, they host people, who are victims of great financial crisis, for six months only. They have two kinds of units:
(1) Individual units and
(2) Family units.
SHN building can be divided into two types of units: A family unit is 400 square feet and an individual unit is 200 square feet. The building can host 20 units in total. Based on experiences from previous years, they have figured out that families can afford \$3,500; however, individuals can contribute to SHN only \$2,000. Edward, David, and Jane hope to serve people in need and also maximize the amount of money that SHN receives from this project to support people in need in other areas.

- How many individuals and families should they host?
- What is the maximum amount of money Edward can raise from this project?

(including some students) tend to explain the issue through the dominant discourse that points at victims rather than the system and structure. This point is one of the premises of neoliberal ideology, according to which concepts such as success, failure, poverty, and unemployment are individual matters that have nothing to do with the sociopolitical and socioeconomic system. Therefore, when it comes to poverty-related issues, students from poor families feel very strongly. They usually feel ashamed. I anticipated that a discussion of this topic might make some students reluctant to join in conversations and feel comfortable exposing themselves. For this reason, I realized that I had to frame the project in such a way that students were able to consider poverty and its related issues as a sociopolitical and socioeconomic problem, instead of merely an individual matter.

Victim blaming is a way to individualize injustice and deflect from a discussion of the public sphere and public good. This individualized ideology

MATHEMATICAL INEQUALITY AND SOCIOECONOMIC INEQUALITY 69

blocks open and transparent discussions, enabling top-down imposition and the manufactured consent of masses (Chomsky & Herman, 2008). This notion aligns with Gramsci's (1971) theorizing of hegemony. What is new here is that under the sway of neoliberalism, the public sphere has contracted in size, resulting, according to Habermas (1975), in a legitimation crisis. Even so, if people are provided with opportunities, they may develop alternative perspectives to look at social and political issues. To achieve this end, in this project I aimed to make small openings in the lives of students where they can connect mathematics to larger societal problems through dialogic pedagogy and collaborative learning.

In my teaching, I had to ensure that this process not become a propagandist approach, about which Freire (2013) cautioned us:

> Propaganda, slogans, myths are the instruments employed by the invader to achieve his objectives….True humanism [and humanizing education], which serves human beings, cannot accept manipulation under any name whatsoever, for humanism there is no path other than dialogue. To engage in dialogue is to be genuine. (p. 101)

Freire (2013) said that liberating teaching can be achieved by genuine dialogue, whereas propaganda and manipulation result in domestication. While improving their mathematical content knowledge, I provided my students with an opportunity to discuss and elaborate on the issues of homelessness, poverty, success, failure, selfishness, generosity, and inequality from different perspectives. They shared their initial thoughts, experiences, and ideas while listening to their peers deconstruct and reconstruct their knowledge and values.

It may seem unusual to align the mathematical concept of inequality with larger sociopolitical problems. Within a traditional education view, it would be very unusual, if not impossible, to have such a discussion in a mathematics classroom. In fact, direct teaching and content-based monologues are the preferred instructional strategies in mathematics classes (Alexander, 2006). However, my study was aimed at praxis: As W. Carr (1995) explained, teachers who only cover content in their classroom are not engaged in educational praxis.

Day 1

Developing egalitarian and collaborative peer interactions was an ongoing part of the entire action research project, and thus part of this project, EUP 4. We began the lesson with a brief discussion about the quality of our group

70 CHAPTER 5

work by reviewing the regulative norms of the collaborative learning process. As Pine (2009) pointed out, "Collaboration is not achieved naturally" (p. 155)— it takes time and practice. Reflecting on our previous projects, the whole class reviewed important factors in collaborative learning and dialogic interactions, raising such features as mutual respect, trust, listening to others, not excluding anybody, welcoming every member to contribute, and non-dominating communication. We continued to strive toward these objectives in subsequent projects, keeping in mind Groundwater-Smith et al.'s (2003) caution that inquiry-based collaboration cannot be achieved overnight—it takes time and continuous effort.

Immediately after the discussion, I handed out the paperwork for the project and had students quietly read the problem. The students then configured their own groups. As groups began working, I circulated around the class to observe peer interactions and assist students as needed. As I was circulating and taking notes, two students in the same group raised their hands:

Jennifer: We are kind of lost...with getting variables.

Tibu: I am confused....We could not figure out the number of unknowns, would that be two or four?

I took a seat in the group and attempted to lead them to use their abilities instead of telling them the answer directly. I was thinking aloud to collaboratively determine the variables, as the following dialogue illustrates:

Me: Ok, let's see....Assuming that Edward accepts one family and one individual, can we write an equation for the total money?

Jennifer: Yes, [she writes it] 3500 + 2000.

Me: Ok, good...What about if he accepted two hosts for each?

Multiple students: Ok, it would be...7000 + 4000.

Me: That is correct....We then could also rewrite as $2 \times 3500 + 2 \times 2000$ [I wrote them down] and reiterate this more like $3 \times 3500 + 3 \times 2000$....Would you agree with that?

Multiple students: Yeah! It makes sense.

Me: We can do the same for the first equation...[I wrote] $1 \times 3500 + 1 \times 2000$....Then what would we think that numbers that increase 1, 2, 3 represents in the problem?

Multiple students: How many rooms are...family and individual units.

MATHEMATICAL INEQUALITY AND SOCIOECONOMIC INEQUALITY 71

Me: Then these are the variables....Let's say x = number of family units, and y = individual units....Great, can you continue from this point?

Tibu: Yes, I guess we can, thanks, Mr. B. [other students nod their heads].

From the excerpts above, I immediately noticed two elements about the quality of group work: First, the students' questioning techniques were much better than before. Second, they seemed to be very careful not to dominate the group. The students seemed to negotiate the question in their group, which was specific about variables as they indicated their confusion if it should be two or four, rather than posing a generic statement like "I don't get this," which is a very common expression used by students when they do not understand subject matter.

With respect to the scaffolding I provided in the excerpt above, Vygotsky's (1978) concept of zone of proximal development (ZPD) focuses not only on peer interaction, but also on the teacher's role. Similar to Freire's (2000) suggestion of a horizontal teacher-student interaction, the ZPD requires the teacher to be an equal partner in the collaborative learning process, transforming the teacher's role from knowledge transmitter to facilitator (Wells, 1999). When the group asked for assistance, I provided scaffolding and let them continue the process on their own instead of telling them the answer. My communication with the group was based on a validity claim instead of a power claim. Dialogic peer interactions and horizontal student-teacher relationships were gradually transforming our classroom into a community of inquiry.

As the group continued working on the project, I followed peer communication in another group. This group did not ask for help; they managed the problem-solving process on their own, as shown in the following:

Nadia: If we maximize something, don't we need a quadratic function?

Nicole: I think so...but some linear inequalities can be optimized. You know, just like we did in exercises last week from the textbook....

Nadia: We need to maximize a function, but what function?

Nicole: Yeah...that is what we need to figure out....We need to write that equation down first....Here we need to add family and individual unit prices [she wrote inequalities down].

Tom: Why did we set inequalities in standard form and put into slope-intercept later?

Nicole: It is easier to graph it in slope-intercept form.

Tom:	Then we should have set it in slope form in first place.

Nicole: I don't know how to set it. It comes easy to set it in standard form when I read the problem....[They wrote all inequalities and the objective function down].

Tom: I wonder if Edward gets money for this work.

Nadia: Maybe some pocket money....Why not?

Nicole: He is doing a community service here; he has a part-time job... you know, in the other project [EUP 2], he got a part-time job.

Once students agreed that the system of inequality corresponds to Edward's story and the objective function, they moved on to calculations and graphing:

Nadia: Looks like our solution area has three edge points...x-intercept is 20 and y is 15.

Nicole: I got the same, but let's plug them in and see how they work....I also got the intercept of two lines.

Tom: Yes, x and y intercepts are correct.

Nadia: Let's evaluate objective function....Tom, can you do it by calculator?

Tom: Yes, $P(x,y)$ is the objective function right? [pointing to their objective function]

Nicole: Oh yeah, that's the equation [she pointed to the equation]....This is going to calculate the total money they could get.

This quotation reveals that each member of the group contributed to the process of collective thinking and working within each other's ZPD. Even though their skills and knowledge vary, they learned from and with each other. Their learning processes align with Cesar's (1998) argument, according to which collaborative learning does not require one peer to be more competent than the others.

I noticed that the students embraced egalitarian peer relationships; they seemed to be sensitive about not dominating each other. The students' reflective journals in EUP 1 indicated their antipathy toward competitive learning in previous experiences in other classes. Egalitarian collaboration in group work has gradually matured. My observation notes in EUP 4 confirm Kohn's (1992) argument against competitive learning. Competition is not necessary to achieve educational success. Excellence can be attained without pitting students against each other, without producing losers and winners, without applying a paradigm of reward and punishment, without having

MATHEMATICAL INEQUALITY AND SOCIOECONOMIC INEQUALITY

students racing against each other. Knowing their work would not be graded, students were still engaged in the collaborative learning process and they enjoyed the process. In fact, they consciously tried to make group work more egalitarian. Collaborative learning helped my students create healthy peer relations and develop self-confidence.

The quotation shows that the students worked on the project enthusiastically. They did not compete with each other; they appeared not to consider their peers as barriers to their success. This lesson structure did not use the reward-punishment paradigm: It did not produce winners and losers. Instead, the process created an egalitarian learning environment where peers stood in solidarity with each other—Flecha's (2000) description of collaborative learning promotes precisely this type of process. This project—without "the use and salience of extrinsic motivators" (Kohn, 1992, p. 221)—promoted a sense of community, and encouraged students to take ownership of their learning.

Day 2

We began the second day by briefly reviewing our activities the day before. We then spent an hour on whole-class discussion and half an hour on journal entries. Taking Edward's story in the context of mathematics in action, the discussion aimed to foster values such as social responsibility, caring about each other, and cooperation in order to counter elements of neoliberal pedagogy such as individualism and competition. I posted topics of the discussion on the board through a PowerPoint:

– What kind of knowledge and skills do you think Edward (and his friends) need to carry out their community voluntary service?

In response to this question, students exchanged ideas and made comments about skills and knowledge oriented toward citizenship responsibility. An excerpt illustrates:

Tom: Edward should know how to label unknowns [in a word problem] first.

Jennifer: He needs to know how to turn sentences into inequalities correctly.

Akil: He's got to know his algebra.

Multiple students: Yeah, obviously...

Selena: He should know how to set the objective function....We struggled with that but figured it out in the end.

Nick: Edward must know stuff like how to graph [linear] inequalities.

Jacob:	Edward's got to know his math, but it is not enough. He also needs collaboration skills to work on this problem with his friends just like we do in this class.
Michael:	I think, before all, he's got to be someone who believes in social responsibility and helping and caring about each other.
Nick:	He volunteered for SHN; I guess we can assume that he is a responsible and helpful guy.

This dialogue reflects the main claim of CME that students need certain values, skills, and attitudes to become critical citizens (Aguilar & Zavaleta, 2012). The students agreed that Edward needed not only mathematics knowledge and skills, but also critical thinking and critical literacy. Their intuitive understanding parallels Gutstein's (2006) distinction between critical and functional literacy:

> A literacy is functional when it serves the reproduction purpose....In contrast, critical literacy means to approach knowledge critically and skeptically....Being critically literate also means to examine one's own and other's lives in relationship to sociopolitical and cultural-historical contexts. (p. 5)

When we apply this distinction to Edward, algebra skills and knowledge (functional literacy) would help him adapt to existing conditions. However, he needed critical mathematical literacy to take the initiative to challenge the existing conditions, for example, helping those in need in his community.

Helping people in need frames the sense of social responsibility of participatory and social justice-based citizenship (Westheimer, 2015). Continuing the discussion in this context, I asked the following questions:

– What do you understand by social responsibility? Do we have any social responsibility?

Tom raised his hand:

> Yes, we all have social responsibilities. For example, the other day, I was walking down the street and saw a couple of banana peels. Well, I picked up the banana peels from the ground and put it in the garbage. What if some dude runs over his car, slips, and crashes his car and dies....That did not happen, because I did my social responsibility....I was a good boy!

Some students responded to Tom's speech with smiley faces. Tom himself had a smile as he said this. I momentarily thought that he was making up this story.

MATHEMATICAL INEQUALITY AND SOCIOECONOMIC INEQUALITY

But I could not find any conceivable reason for him to make it up. Regardless of my momentary skepticism, he explained what social responsibility might mean with a very simple and effective example. It is about caring about each other in a community.

From a neoliberal perspective, however, responsibility can be defined only at the individual and family levels (Ventura, 2012). Acknowledging the notion of social responsibility, Tom's comment challenged the neoliberal standpoint. Therefore, I transitioned from the question of responsibility to the question of choice between a competitive society and a collaborative society. I posed a question to invite students to share their ideas and reflect on their experiences. As I asked it, I was aware that the idea of competition is continuously imposed in the U.S., as Kohn (1992) indicated:

> [In the U.S.] we are systematically socialized to compete—and want to compete—and then the results are cited as evidence of competition's inevitability....The message that competition is appropriate, desirable, required, and even unavoidable is drummed into us from nursery school: it is the subtext of every lesson. (p. 25)

My next question, therefore, focused on the desirability of competition:

– What kind of social environment would you wish to live in, cooperative or competitive? Would you want our learning activities be cooperative or competitive?

Andrew:	Because of the Manhattan Project's competition, nuclear power was invented. I think competitive society is more productive.
Tori:	Nuclear power could be invented by cooperation as well. I personally believe that competition overall produces a hostile environment....I mean, if the U.S. and Russia [had] cooperated, the world would have been a better place.
Bin:	Frankly, I believe that collaboration is better than competition. You know...I enjoy studying together in our class, I think we all do, but there are some other things that I would like to do on my own, and I don't think that contradicts with the first.
Jennifer:	I like how we study in this class. Our chemistry class starts like, there is "do-now" question on the board, like, you know...whoever gets the answer first gets bonus credit. Sometimes, I mean, even when I know the answer, I don't want to raise my hand.

The excerpt relates to Kohn's (1992) claim that competition is ideologically imposed from the top down and is perceived almost as a national religion. Based on my observations, student journals, and my reflective journal, the hegemony of competitive learning can be countered. EUP 4 revealed that small openings can be created for students to learn without competition.

As I was preparing to ask if they remembered a sports game they had lost, Ceylan added another dimension to the discussion:

Ceylan: I really would like to live in a world [where] people help each other and collaborate to solve their problems; they play games for just fun. But this is not the world we live in....We have to compete one way or another. Even in our school, next period we go to another class and Mrs. [X] will say, "There is a do-now question on the board. Whoever gets the answer first will receive extra credit." Welcome to the competition!

Nick: That is not only Mrs. [X]; that happens in many other classes too.

Bin: It is not only school. My father says that the business world is a place where dog eats dog.

Me: Unfortunately, our world is formed by the competitive perspective, and we have to survive in this world. But always remember that the competitive approach is a human-made decision; it does not come from god. And, therefore, it can be changed. If we think cooperation is more humanizing, you, young citizens, have a power to challenge the existing system that glorifies competition. We began this change in our classroom, and we can continue in other parts of our life.

This was a challenging moment for me in relation to critical pedagogy. It was not easy to respond to Ceylan's comment; this aligns with McLaren and Leonard's (1993) point that "critical pedagogy must serve as a form of critique and also a referent for hope" (p. 69). In other words, the practice of critical pedagogy runs the risk of generating hopelessness. In that sense, I attempted to promote the idea that as much as we have to survive in a given system, we should try to transform it if we are unhappy with it.

As Edward's story involved a homeless shelter, I invited my class to reflect on the notions of success and failure. Neoliberal ideology defines success and failure as individual matters. I posted the following question on the board:

– Do you consider success or failure as an individual or social matter? Explain your reasoning.

MATHEMATICAL INEQUALITY AND SOCIOECONOMIC INEQUALITY 77

Me: Let's evaluate this question a bit more. We sometimes fail at something and succeed at another. That is how life goes on. But when it comes to explaining reasons behind our failure or success, you may have different thoughts…

Tibu: My dad was laid off from Amazon and could not find another job for six months. We had a really hard time.…He got a job in a grocery store with minimum wage. Was this my dad's mistake? He is a very hard-working man.

Andrew: Yeah, but what about lazy people who just give up?

Jeff: People lose their job because of the bad economy.…It would be unfair to blame individuals in this case. You know, you get wet when it rains, they say, why don't you have umbrella? Well, if it rains heavy, you get wet even with a nice umbrella.

Darryl: Success and failure, I think, is an individual matter. If you have a rich father and you can have everything you need and want.…Let's say, you don't study and then fail all your classes; that is definitely an individual matter. But if you didn't have dinner last night as your dad lost his job or you live in foster care, and you failed your classes…I would not blame you for this.

These comments went against my anticipations. I was expecting that most students would tend to think of success and failure more as individual matters, as that is the dominant discourse in the U.S. The students seemed to challenge the neoliberal assessment of failure and success with reasonable arguments. Specifically, Darryl's example of a student's failure demonstrated situations under which failure could be either an individual or a sociopolitical and socioeconomic matter.

As I gazed at the clock on the wall, I saw we had 28 minutes left for journal entries. I posted the questions below through the projector and handed out the journals. For the rest of the period, the students made their journal entries for the project.

Students' Journals
I posted prompt questions on the board:

– Briefly describe your group's collaboration story.

The students' journal entries indicated that they genuinely exposed their feelings, thoughts, and ideas. For example, Taylor articulated his joy of applying mathematics to a real-life situation:

> We did really well on this project. We did not get sidetracked at all. It was enjoyable; we did not agree on the objective function at first, but we talked about it for a while and figured it out. Actually, I felt like I learned it with understanding this time. Also I began looking at things from mathematical perspective that I have never done before like we did in this project, using systems of linear inequalities for community service problem....Math is used pretty much for everything.

Taylor's comment shows that in collaborative learning, students can learn subject matter "with understanding" as they learn with and from each other through dialogue. This finding is consistent with educational theories arguing that fellow students can widen your horizons much better than you can do on your own (Skovsmose & Alrø, 2004; Vygotsky, 1978; Wells, 1999). Ever since the first project, the students emphasized dialogue and collaboration as we had agreed on improving the quality of group work. Bin's comment was unique, as he consciously considered the limitations and possibilities of collaboration:

> The quality of our collaboration is improving, but I would not say it is perfect. When I first saw the problem, I knew what to do, but I let my friends speak their ideas before I shared mine....Although everyone made a contribution, they have remained silent time to time....I pushed more and more to hear what they think, but I also tried not to be dominant as well. I would have liked to exchange ideas more enthusiastically and ask questions to each other more.

Bin's entry indicates that he consciously avoided being dominant in the group. Although he immediately knew the mathematical aspect of the word problem, he was considerate enough to include his classmates in the process by thinking that his peers' thoughts would enrich the overall quality of their work. Bin's entry reveals that empathy plays an important role in egalitarian peer interactions. His approach to his classmates was not only dialogic, but also empathetic. Considering Bin's comment in light of other students' journal entries, I concluded that Habermas's ideal speech situation could be attained. I will return to this point.

The next question invited the students to share their experiences and ideas about whether there need to be institutions to assist people in need:

– Do you think that societies should have a safety net?[1]

This question relates to whether success or failure is an individual or a socioeconomic and sociopolitical matter. If it is not an individual matter, then how do we approach it? Jeff wrote:

Societies definitely should have a safety net, due to the fact that some people get placed in a situation of disadvantage just from sheer bad luck....Safety net should exist to aid people in growing back into a state of independence since even hard-working and good people may get down on their luck and need help.

Jeff implied that within the existing system, everyone, including hard-working people, can be a victim; therefore, a society should have institutions to assist citizens in need. Jennifer added another layer:

I think we should have a safety net, but I don't think that it is a permanent solution....As long as there is an unfair system, there will be people in need. The only people, who care about these in need, are people who volunteer for non-profit organizations....The government is not genuinely interested in helping people who struggle to meet basic needs.

She argues that one needs to focus on the root causes of socioeconomic problems if one really wants to solve these problems. She also points out that a safety net is the responsibility of government; citizens should not be at the mercy of non-profit organizations or charities.

Michael proposed a similar approach:

This safety net thing is meaningless; it does not solve the problem. If people work for six days a week and twelve hours a day, and still can't survive and need food stamps....Well, we need to look elsewhere to get the real problem here. These smart gentlemen better stop blaming lazy people and explain why hard-working people fail to make their bread. We should work for a more equal society.

From Michael's point of view, our efforts need to focus on making our society more equal and just. If that were achieved, social programs such as the social safety net would not be needed.

Tori agreed on the necessity of a safety net by pointing to unfair conditions resulting from a market-driven society: "There is so much competition that some people like elderly, veterans, disabled are left on the side and have little or no option. That is why we need safety net." Jacob represented the necessity of a safety net as an indicator of caring about each other. From his point of view, a group of people turns into a community when they care about each other: "Yes, we need a safety net, we need to care about each other and help those who are in need. This is what being a community means."

The students' comments showed that they envisioned a society where people care about each other. Without exception, all students disapproved of the neoliberal standpoint, according to which individuals are responsible for themselves and have no right to expect help from society (Ventura, 2012). The next questions concerned Edward's possible motivation:

– Why do you think that Edward volunteered for SHN organization? What could be his motivations? How and why?

With this question, I aimed to help students to unpack two main philosophies behind community volunteer services and further elaborate on them. The first is that community service is equated with being nice and helpful; it aims to help people in need, but never questions why people are in need in the first place. The second philosophy politicizes the process: While providing help for people in need, it questions the root causes that put people in need.

To respond to this question, students reflected on their life experiences to come up with educated guesses. For example, Derek wrote, "Perhaps Edward himself was once was homeless or struggled financially and decided to volunteer for SHN." Darryl responded, "Edward wanted to help people in his community who needed it. He probably felt socially obligated to stand up for those who are not as lucky as himself." Nicole wrote, "Perhaps Edward volunteered for SHN to improve his resume as universities require community involvement....Or maybe he was thinking that it was a social responsibility...or he was a very caring individual." Ryan stated, "As we know he was looking for a job earlier, he may have realized that people are better off when they help each other in hard times."

In their answers, the students actually articulated why they would volunteer for SHN from their own point of view. The students indicated that their sense of social responsibility, empathy, and caring about each other and their community were emergent. These qualities are what Westheimer (2015) considered essential for justice-based and participatory citizenship. The next question invited students to reflect on their experiences with notions of generosity and selfishness:

– Do you think that people are caring or selfish by their nature? Or people's actions should be evaluated within given social-economic system? Give examples from your life experiences.

Peer collaboration is central to our projects. However, if students tend to think that people are selfish by nature, why would they genuinely participate in a collaborative learning process that requires caring about each other and

MATHEMATICAL INEQUALITY AND SOCIOECONOMIC INEQUALITY 81

negates selfishness? This implies that collaboration can be achieved if students embrace values and attitudes such as collaboration, solidarity, and caring about each other.

The students provided thoughtful responses to the question by reflecting on their life experiences both in and out of the classroom. Christian wrote:

> People are caring or selfish not by nature but by experiences....How others treat them in a community forces them to be one way or another. For example, when my buddies help me with my math in our group, I would like to help them any way that I can.

Christian's comment reveals that the socioeconomic and sociopolitical structure urges people to act in certain ways. A collaborative and dialogic learning environment motivated Christian to help his classmates. In education, if a learning task is designed to be competitive, students might seem to be competitive. However, this does not mean that competition is inherent in human beings.

Jennifer specified that people become selfish as they try to survive in a competitive society: "I don't think people are selfish by nature, but a winner-takes-all society makes people selfish; in order to be the best, you need to defeat rivals. Sometimes these rivals are your neighbors and sometimes your friends." She implies that survival in a market-driven society requires a selfish attitude. Otherwise, in agreement with Kohn (1992), Jennifer does not consider people inherently selfish. The students' comments clearly indicated their understanding that people's actions cannot be separated from the social, economic, and political system in which they live.

The next question invited students to identify certain issues either as individual or social problems:

– Do you think homelessness is an individual problem or a socioeconomic and political problem? Why?

Nadia wrote, "Homelessness in the end is a social problem. Nobody enjoys being homeless or [being] in a difficult situation." Her conclusion is based on the premise that if homelessness is not a choice of the individual, then it has something to do with the socioeconomic and sociopolitical system. Cindy supported her conclusion with concrete examples:

> It doesn't make sense to me when people say that homeless people are homeless because they are lazy. What if you lose your job or get fired, or you are disabled or sick? It is same for people with drug addiction. They

need guidance and medical help to get better, but they are thrown into jail instead in our country....There are homeless orphans on the street. If society or the government helps them, they may be able to stand on their feet; otherwise they will always be on the street.

In Cindy's view, homelessness is a social problem and so is its solution; it might result from being laid off or fired, sick, or addicted to drugs. If we as a society stand in solidarity, these problems can be effectively solved. Darryl seemed to agree: "If you are lazy, [homelessness] is a personal problem, but most people are not so. They may have been laid off or have had bad luck and circumstances that have prevented their success." Nick argued that homelessness cannot simply be reduced to an individual matter:

> I think that homelessness is a political and socioeconomic problem because no one would like to be homeless. Something had to happen to them, such as being denied a job or failing to graduate, or any other reason that caused a setback for them. And we can't blame all of that on them.

Neoliberal ideology rejects thinking of problems at a social level. Instead, it considers one's failure, success, misery, or happiness as individual matters rather than social, economic, and political problems (Ventura, 2012). However, even though they were not able to use the word "neoliberalism," the students' comments show that they challenge basic tenets of neoliberal ideology.

The next question encouraged the students to provide an overall reflection on this project:

– This is our fourth project; overall do you feel empowered in a sense that you can apply your math knowledge and skills to real-world problems? Please be as specific as possible.

Students' responses indicated that they took ownership of the collaborative and dialogic learning process; they were empowered, as they could use their mathematics skills and knowledge outside of class. They also critically evaluated the joint objectives oriented toward creating a community of collaboratively learning mathematics. For example, Jeff noticed improvements in peer dialogue:

> I feel that this project, I felt for the first time that I learned when my friends answered my questions as well as listening to my classmates asking or explaining each other's questions and confirmations....Also

I enjoyed our classroom discussion that connected Edward's Community Volunteer Service to social problems.

It is evident in Jeff's statement that students experienced CME by dialogic learning and by connecting mathematics to larger social, political, and economic problems. Nicole articulated achievements up to this point:

As we do more projects, our collaboration and dialogue got better; no one dominates, everybody contributes, and we listen to each other to solve the problem, and we all take pride [in our] final work....I also realized that skills I developed in our class for collaboration helps me for other classes as well.

Nick shared how our projects gradually changed his perspective:

As we did projects, I feel I am much more empowered on looking at things in life from a mathematical perspective....Until I got in this class, math was for me to study for the test subject. But these projects opened my eyes....Edward used his math knowledge to figure out his best salary and now he uses math to raise maximum money to serve people in his community. This is a very unusual story for me.

The comments suggest that EUPs were successful in fulfilling one of the fundamental goals of CME: to transform students' "foreground" (Skovsmose, 2011). Students clearly recognized their changed and changing ways of thinking about mathematics, doing mathematics, and reflecting on mathematics. Aligning with Kemmis, McTaggart, and Nixon's (2014) description of CPAR, my students and I critically examined learning practices in the class and jointly attempted to transform the class into a community of learners. Darryl outlined how he benefited from collaborative and dialogic learning:

I do feel empowered. This class has been very fun for me because solving real-world problems helps me to understand concepts and how they apply to real-world situations. Working with my classmates made everything easier and has allowed me to make connections to more complex mathematical concepts and theories.

These comments show that the whole class made tangible progress toward being a community of collaborative learners. Having reflected on the previous EUPs, I realized that there is a dialectical connection between practicing CME and the classroom being a community of learners.

84   CHAPTER 5

Community Service: Charity or Solidarity

One major challenge of this action research project was to decide and substantiate themes of EUPs. When I considered community service as the sociopolitical theme of EUP 4, I faced the same challenge. I realized that it is easy to fall in neoliberal pitfalls as neoliberal ideology and pedagogy also promotes community service volunteer projects. What I learned from this project would guide me if I were to do similar classroom projects in the future; it would also provide a vision for critical pedagogy teachers who wish to be engaged in "counterhegemonic" teaching and learning practices.

Two main lessons emerged. First, a project that involves themes of community service should have two objectives. The project should allow students to notice that helping those in need is a virtue that an active/ critical citizen should possess. Neoliberal pedagogy implicitly promotes individualism and selfishness. This should be problematized and countered. However, a project should also allow students to ask, "Why are these people in need first place?" Counterhegemonic learning and teaching processes should urge students to identify root causes embedded in the socioeconomic and sociopolitical structure that produce and reproduce poverty, misery, and inequality. Without radically questioning root causes of socioeconomic inequality, an educational process in the context of community service may end up promoting neoliberal pedagogy. In Freire's term, this process would promote a sense of "false generosity."

The second lesson is that "who is helping whom and why" should be clear in order to distinguish solidarity from charity. The process of helping those in need or volunteering should not generate or reinforce any power relation. Helping each other should occur among people who have comparable socioeconomic power. In this way, helping each other in a society can nurture citizens with caring personalities and cultivate social solidarity. But if help comes from power elites or the ruling class, rich people's charity may do much more harm than good in poor people's life-worlds. Today in the U.S. there are many nongovernmental organization (NGO) charities run by power elites that provide "help" for those in need. An educational process inspired by critical pedagogy should enable students to understand ways in which rich people's charity is actually harmful because it implicitly normalizes power relations and justifies existing inequalities and poverty. Rich people's charity aims to manipulate class consciousness among working-class people and generates the idea that inequality and poverty are inevitable; it creates a perception that living in a state of inequality and poverty is some people's destiny.

Concluding Discussion of EUP 4

Inquiry-based collaborative learning has been central to this project as well as to previous projects. As a whole class we negotiated potential obstacles to collaboration and agreed on eliminating these obstacles collectively. For example, in EUP 1, students expressed concern about "self-assigned leaders" in group work, and proposed a more egalitarian peer relationship. We set certain goals to make our communication more dialogic and collaboration more egalitarian. For each EUP, the students continued reflecting on these overarching goals as well as specific objectives within each EUP. Having triangulated my observations of whole-class discussions and peer interactions, students' journals, and my reflective journal, this project has led me to five main conclusions.

First, when the students worked on Edward's story in groups, they learned from and with each other to construct knowledge collectively. Based on Vygotsky's (1978) conception of peer interaction as a ZPD, we know that through collaboration, students can achieve much more than they could on their own. My observations also confirm that for a ZPD to be realized, a more competent peer within a group is not necessary. Once dialogic peer interaction is achieved, students can work in each other's ZPDs, as suggested by Cesar (1998). Vygotsky's notion of ZPD intersects with Freire's "horizontal" student-teacher relationship (Freire, 2000; Wells, 1999). When students asked for help, I provided scaffolding as just another member of the group, instead of posing as the ultimate source of knowledge. This is not to deny that, as the classroom teacher, I had authority. However, in light of CME and critical pedagogy, I used my authority to establish dialogic (horizontal) student-teacher relationships and a facilitative learning environment.

Second, the students stated their preference for collaborative over competitive learning; they said they learned better if they studied in groups in which peer communication was dialogic and peer interactions were egalitarian. This confirms Flecha's (2000) argument that dialogic learning requires egalitarian peer relationships, non-authoritarian teacher-student relationships, and reflexive dialogue. Inquiry-based learning and dialogic interactions transformed the classroom into a community of inquiry. Being a community of collective inquiry creates the material conditions necessary for Habermas's (1987, 1990) theory of the *ideal speech situation*. When applied to educational settings, these material conditions are as follows: (1) the students included their peers in a collaborative learning process; (2) each student was free to question any argument that came either from the teacher or from peers; (3) students were free to express their arguments; (4) no external force was imposed on the process—all decisions were reached through noncoercive arguments. As a result, knowledge was constructed based on validity claims, not power claims.

Third, this process could not be fully constructed by a mechanical exchange of ideas and thoughts within regulative norms. With respect to this question, Freire (2000) introduced another dimension of educational process: love[2] and hope. From Freire's point of view, dialogue requires love, humility, and faith:

> Founding itself upon love, humility, and faith, dialogue is a horizontal relationship of which mutual trust between the participants is the logical consequence....Nor yet can dialogue exist without hope. Hope is rooted in men's incompletion....The dehumanization resulting from an unjust order is not a cause for despair but for hope. (p. 91)

My observations and students' reflections guided me to conclude that the ideal speech situation in an educational setting could not fully be realized in the absence of love and hope. Egalitarian peer interactions and horizontal student-teacher relations require more than an exchange of cold arguments. As indicated earlier in connection with Bin's journal entry, I realized that empathy plays a significant role in peer and teacher-student interactions. Sometimes I would teach the students skills, knowledge, and values over and over again until I succeeded. This was not a mechanical process. Freire (2000) points out that a critical pedagogy cannot be materialized without love and hope, and it was during EUP 4 that I realized that without hope and love, it would be easy to become discouraged.

More importantly, I observed the students genuinely striving to explain their arguments for their classmates and respectfully listening to their peers' views without trying to dominate each other. As suggested by Freire (2000) and hooks (2003), the students interacted through empathy and love. It can be concluded that creating an ideal speech situation in the classroom can be achieved not only through better arguments, but also through empathy, as is evident in Bin's journal entry. This finding confirms the philosophical foundation of this study—the complementary ideas of Freire and Habermas.

Fourth, as I answered the students' mathematics questions, I drew on the validity of mathematical axioms and algebraic properties without playing the role of an external authority who was imposing arbitrary rules or laws. While answering their questions, I did not move to the next step unless the students were convinced by the justification I provided. In other words, my teaching mathematics was not based on a power claim but on a validity claim.

This approach to teaching mathematics resembles that of Almeida (2010), who compared teaching mathematics through dialogic pedagogy with the dynamics of democracy as a political system. Democracy refers to the Constitution whereas mathematics refers to theorems, properties, and axioms to establish the validity of an argument.

Fifth, in the context of Edward's story, the students negotiated, formed, and reformed values that are necessary to become critical citizens. The students articulated that they preferred collaboration to competition; they spoke in favor of a society where people care about each other in solidarity; they supported egalitarian peer interactions; they established a dialectic connection between the individual person and society (life-world and system). As we engaged mathematics through inquiry-based collaborative learning and dialogic pedagogy, students had opportunities to experience democracy in the form of dialogue and internalize democratic culture. In other words, the students embraced values that challenged the hegemonic values of neoliberalism.

Notes

1 The safety net refers to the welfare state and institutions that are in place to provide citizens in need with help.
2 Love in this context should not be confused with romantic love; hooks (2003) described this kind "as a combination of care, commitment, knowledge, responsibility, respect and trust" (p. 131).

CHAPTER 6

Student Loan Crises

In this chapter, I analyze the ways high school mathematics can be taught and learned through dialogic pedagogy. I argue that the ideas of Freire and Habermas can inspire dialogic teaching of mathematics. Authoritarian (non-dialogical) teaching is the dominant approach in the traditional education system; thus transforming a math classroom to a dialogic classroom takes time and systematic effort. I also argue that the dialogic teaching of mathematics cannot be materialized in the absence of inquiry-based collaborative learning.

The chapter presents findings from EUP 5 that contextualize exponential function-equations into the student loan debt (SLD) crisis. The students believe that education is a human right and thus should be free of tuition. They criticize higher education in the U.S. for not being affordable for the majority of students; SLD, they argue, steals their future. The students make short- and long-term proposals to remedy the situation.

Planning and Objectives

The SLD crisis has long been a major issue for students who come from working-class families in the U.S. Resulting from the neoliberal policies over the last thirty years, the SLD crisis is getting worse.[1] Since 1980, the average income of working-class people in the U.S. has significantly decreased while the cost of living has increased (Hayase, 2013; Torbat, 2008; West & Smiley, 2012). This increase includes college tuition and student loan interest rates (Brown, 2015). Employment opportunities for college graduates have diminished, which makes paying back debts more difficult than ever for new graduates (Blacker, 2013).

Although the SLD issue directly affects students in high school, they usually have no opportunity to understand and critically reflect on this matter. There is a career center in our school where students can get help with college applications, student loans, and scholarship problems, it implicitly invites students to adapt to existing conditions without any criticism. That brings us back to the hegemony of neoliberal ideology. In the neoliberal view, issues such as SLD are individual, not social matters. I therefore developed another story about Edward, a senior student in Liberty High School. Previously, Edward was a part-time worker and a volunteer in a community service. Now he's getting ready for college.

© KONINKLIJKE BRILL NV, LEIDEN, 2019 | DOI:10.1163/9789004390232_006

STUDENT LOAN CRISES   89

CME should provide students with opportunities to connect their life in the classroom to the larger social, political, and economic systems in which they live (Ernest, 2002a; Gutstein, 2006; Skovsmose, 2011). In this vein, inspired by Freire, Shor (1993) said, "In the liberating classroom, teachers pose problems derived from students' life, social issues, and academic subjects, in a mutually created dialogue" (p. 25). In light of these critical educational theories, the SLD issue exemplifies a problem derived from the lives of my students.

I designed the paperwork so that each group would engage in an inquiry-based collaboration and dialogue to envision Edward's life after college and reflect on it. The first page of the project was as follows:

EUP 5

Student Loan Crisis: Life after High School

Edward is a senior student at Liberty High School. He is planning to study *engineering* (this is your group's decision). He is going to need a loan for his college tuition (4 years) and for some of his expenses while he is in college. Assuming that his loan comes with 6% annual interest, soon after graduation, Edward will start paying back the loan as a monthly-fixed amounts and he is supposed to pay it back in 10 years. Edward is aware of the student loan debt crisis in the U.S.; therefore, he wants to take educated and conscious steps toward his life in college and after college. From this point, through his mathematics knowledge and skills, he wants to model different scenarios and be ready for some of the risks and possibilities he may encounter. He is going to do mathematics for:

1. The amount of the loan he is to apply for and the total money he is supposed to pay back in ten years.
2. The amount of his monthly payment in connection with his other expenditures such as rent, food, and other monthly bills.
3. Ideally, he wants his monthly payment to be between 10% and 20% of his monthly income when he gets a job. Based on his loan payment, what kind of job should he hope to get?
4. He wants to make some predictions about how his financial situation and social life will look like while he will be paying back his loan.
5. Edward also wants to reflect on the student loan crisis not only as an individual, but also as a responsible and critical citizen.

The project was designed as an open-ended word problem that aimed to provide students with a "landscape of investigation," as suggested by

Skovsmose (2011), through which the students could develop multiple scenarios for the life-world of a senior student who has to deal with the SLD issue.

The students were to consider Edward's life-world from a sociopolitical, socioeconomic, and cultural perspective and critically reflect on his life. Each group was required to reach a consensus on multiple points, such as the initial amount of the loan, Edward's possible expenses, lifestyle, employment situation, and payment plan. Furthermore, through mathematical calculations, each group was expected to come up with several different scenarios, such as Edward has a decent-paying job, a minimum-wage job, or no job at all. In comparison to the previous projects, this one required a high level of collaboration and dialogue in order to reach consensus on each point.

Day 1

Prior to this lesson, we spent four periods on processing exponential properties, functions, and equations. But we had not covered compound interest calculation, which this project required. Therefore, we began the project by learning the compound interest calculation formula. I aimed to construct the formula with my students to ensure they learned it through active participation and with understanding. I had a lesson plan to follow, but I did all the calculations with my students as our dialogue progressed. I projected an example on the board to begin the lesson:

Me: Alright....If you guys are ready, before working on our student loan project, we have to develop a mathematical model to calculate compound interest....Let's say, we invest $1,000 in a bank with 3% annual interest for five years. We will figure the total amount of money we are supposed to get back at the end of the fifth year....How much money will be added at the end of the first year? [The question was directed to the whole class.]

Multiple students: Just multiply $1,000 with 0.03...it is $30.

Me: That will calculate how much is to be added to the starting amount. If we invested for a year, how much would we get back? [An office assistant stepped into the class to give some letters to a student. I asked her to come back five minutes before class ends—I try to avoid disruptions.]

Nicole: You know, just add 30 to 1000....[Multiple voices] It is $1030.

Me: Correct....Ok, then, let's move on to calculate the amount for the second year. Should we start with $1030 or the original

STUDENT LOAN CRISES

> amount, which was $1000? [The question was directed to the whole class.]

Multiple students: It is $1030.

However, Nadia, who was sitting in the back-left corner table, said, "Original amount." As I was getting ready to begin a dialogue with her, Tom responded: "If we take the original amount each year, it would just be $30 extra per year after five years. It doesn't make sense." Ceylan joined this discussion: "I think you would be right if it was 3% rate without accumulation, but it says 3% annually for five years." Nick joined in: "If I earned $30 for the first year, then I have $1030. It is my original money, not $1000." At this point, a discussion between the students and me triggered peer-to-peer communication and transformed the classroom into a dialogic sphere where collaborative learning was taking place. Nadia seemed to be convinced by Ceylan's and Nick's explanation.

The peer communication revealed that the students seemed to understand each other's point of view. They felt comfortable expressing their ideas as they had already developed a sense of belonging to a community of collaborative learners. This finding reveals that dialogic and collaborative peer interactions in the classroom generate a sense of community.

It would be very difficult, if not impossible, to create this learning environment in a classroom where communication takes the form of a monologue between the teacher and an individual student. Even though I was teaching the whole class, the students worked in multiple zones of each other's ZPDs because the emergent dialogic classroom allowed this to happen. Collaborative learning in EUPs created a classroom atmosphere where the students perceived their classmates as members of their community of mathematics learners; therefore, they appeared to respect and trust each other's friendly intention.

For example, because learning in our class was a collaborative activity, Nadia seemed to appreciate Ceylan's response. In a classroom where learning is competitive, such egalitarian peer dialogue would be very unusual—the dialogue between Nadia and Ceylan could be seen as a race between two rivals. Perhaps Nadia would not have responded to my question in the first place, as making a mistake or giving the wrong answer is equivalent to being a loser in the competitive learning environment. The preceding excerpt signaled how central an egalitarian community is to sustaining collaborative and dialogic learning.

The next segment shows how a mathematical formula may be introduced in a dialogic, as opposed to authoritative, form of teaching:

Me:	Ok....If we all agree on this, let's calculate it for the second year. [Writing on the board:] $1030 × 0.03 = $30.9, and if we add this to the first year amount, $1030 + $30.9 = $1060.9. Let's do this for all five years; you help me with calculations with your calculators. [Writing on the board, several students made calculations out loud, and I wrote them on the board.]

At this point, I noticed that participating in creating the table above helped students remain engaged and be an integral part of the knowledge construction process. Tom noticed that the final outcome in the third row was incorrect. Jennifer indicated the same mistake. I corrected it. This shows that the students were actively participating in the collective meaning-making process. Once the table was complete, I carried on talking with the whole class:

Me:	[To whole class:] Can we see any patterns here?
Lexus:	Looks like each time we multiply by 0.03.
Me:	Do we all agree with that?
MSR:	Yep! Yes, we do...
Me:	All right then....Is there any other pattern we can locate? [A couple of students raised their hands.]
Me:	Yes, Nick?
Nick:	I think each year we start with the last year's amount.

Multiple students: Yep! That is right.

Me:	Perhaps we can reorganize this table in a slightly different form to make these patterns a little more visible by factoring the expression. As we multiply each year's amount by 0.03, we can rewrite them as below:

[I write them on the second board, at the back of the class. Multiple students did calculations with calculators to double check the outcomes for each year.]

Year 1 $1000 (1 + 0.03) = \$1030$

Year 2 $1000 (1 + 0.03)(1 + 0.03) = \$1060.9$

Year 3 $1000 (1 + 0.03)(1 + 0.03)(1 + 0.03) = \$1092.727$

Year 4 $1000 (1 + 0.03)(1 + 0.03)(1 + 0.03)(1+0.03) = \$1125.50881$

Year 5 $1000 (1 + 0.03)(1 + 0.03)(1 + 0.03)(1+0.03) = \$1159.274074$

Me:	Ok....Please don't lose your focus...[At that moment, a delivery truck was backing up to the school's kitchen entry, which was

STUDENT LOAN CRISES

right next to our class; the truck's noise distracted the students.] Now our calculation is complete, can you please take a look at the table one more time. [Silence for a while.] Would it be possible for us to generalize this pattern and turn it into a formula?

Multiple students: Maybe....Yes...let's try...

Me: [To whole class:] In this table, from one year to the next, what is it that seems changing?

            [No responses for a short while...the students are having a thinking/processing time.]

Jennifer: Each year we add another parenthesis.

Me: Do you mean that we add another multiplication of 1.03?

Jennifer: Oh, yeah! That is what I meant.

Here I exercised caution to ensure that all students agreed before I made the transition to the next step and attempted to justify my points with algebraic axioms and theorems. I was modeling my explanation on Almeida (2010), who argued that drawing on axioms and theorems is a way of democratically structuring the teaching of mathematics.

Me: If we all agree, then the year can be a variable in our formula.... We may improve this table to catch this pattern a little better by using the properties of exponential expressions. [Writing on the same board, but on the right side of it:] Instead of multiplying them by years, we can express it exponentially. Would you please help me with calculations with your calculators?

            [Several students did calculations and I wrote them on the board.]

            Year 1 = 1000 + $(1000 \times 0.03) = 1000 (1 + 0.03)^1 = \$1030$
            Year 2 = $1000 (1 + 0.03)^2 = \$1060.9$
            Year 3 = $1000 (1 + 0.03)^3 = \$1092.727$
            Year 4 = $1000 (1 + 0.03)^4 = \$1125.50881$
            Year 5 = $1000 (1 + 0.03)^5 = \$1159.274074$

Me: Alright....We may be able to generalize these patterns anytime soon. Do these outcomes match with the other table?

Multiple students: [Silent for a while.] Yep....Right column is exactly the same.

Me:	Then we can generalize these patterns and set a formula. We all agreed that the year (time) is a variable; let's label the year as t, as it stands for time.

Multiple students: Ok....Let's go for it.

Me:	In our example, the interest rate was 3%, but it could be different. Should we take the interest rate as a variable as well?
Nicole:	But it doesn't change in the table.
	[Another student in the class responded before me:]
Noah:	But we may have another example with a different rate, you know, just like credit card rates....Each one is different. We can't put 3% for every example.
Nicole:	Got it...It doesn't change in this example, but it might change for others.

The excerpt shows that the students were actively following and contributing to construct a formula collectively. I realized that this way of teaching made me think of myself as more of a facilitator in a dialogic classroom than an authoritarian teacher; I felt empowered. We as a whole class continued constructing the formula:

Me:	Let's call the interest rate r...by the way, different textbooks mark them differently. It doesn't have to be r everywhere. Does inside of parenthesis change for each year?

Multiple students: No....Looks the same.

Me:	Ok, then we can simplify it as 1.03, and the starting amount of money also remains the same in the table. Then we can generalize the entire table through variables in one expression [I wrote the formula under the table on the board.]
	\$Final amount = \$Original amount $\times (1 + r)^t$, where t stands for year, r stands for interest rate
Me:	Can we verify the numbers in the table with this formula to make sure that it matches with our long calculations?
Taylor:	I did the fourth year, it seems correct.
Jacob:	I just did the third...that is true as well.
Me:	[To whole class:] Let's verify all the other years with this formula.

STUDENT LOAN CRISES

Multiple students: Yep...it works.

Me: Then, from now on, whenever we need to calculate compounded yearly interest, we can use the formula we just developed.

As shown in the excerpt, I attempted to introduce the new mathematics topic in a dialogic form. I constructed the formula together with my students. I justified my instruction through mathematical concepts and encouraged students to participate in the process of developing the exponential model. The students actively contributed by verifying the equations' numerical accuracy and algebraic coherence. In other words, I facilitated the process but did not continue my teaching without consent of the students.

Science and mathematics teachers usually take an authoritarian (nondialogic) approach when they introduce a new theorem, concept, or topic (Alexander, 2005; Mortimer & Scott, 2003). Traditionally, mathematics teachers are not supposed to spend this length of time to teach a formula; it is considered time-consuming and not "effective." The authoritarian approach would be to directly provide the formula on the board, followed by several repetitive, skill-drill examples, so that students acquire procedural knowledge of this calculation.

Skovsmose (1994) called this approach to mathematics education the "exercise paradigm"; Freire (2000) conceptualized it as the "banking" concept. According to Skovsmose (1994, 2011), this approach creates a comfort zone for students in mathematics classrooms and thereby makes students bored; they lose motivation for learning. Emphasizing its sociopolitical implications, Freire argued that the banking concept is anti-dialogical: it produces authority-dependent personalities. Establishing and sustaining a democracy, on the other hand, requires active and critical citizens who have communicative competencies and are able to hold authorities accountable (Giroux, 2012). I could have followed the authoritarian path to introduce this formula, which would have taken less time and effort on my side. Instead, I preferred the dialogic approach as I aimed to practice CME, which emphasizes dialogic and collaborative teaching and learning of mathematics to promote critical mathematical literacy. The extract from my whole-class teaching reveals that the dialogue was not simply a means to teach the formula: Dialogue was also an end in itself rather than an instrument for something else.

I co-constructed knowledge—the interest formula—with my students. Even though we did not invent the formula, through dialogue and collaboration, we brought it to life. With my facilitative instruction and students' active partici-pation, the exponential model of interest calculation became meaningful for us. Drawing on Freire's ideas, Shor (1993) pointed out that liberating teaching invites students to active participation; it rejects rote memorization and the transmitting style of education that results in the alienation of both students and teachers:

> Teaching and learning are human experiences with profound consequences. Education is not reducible to a mechanical method of instruction. Learning is not a quantity of information to be memorized or a package of skills to be transferred to students. Classrooms die as intellectual centers when they become delivery systems for lifeless bodies of knowledge. (p. 25)

In contrast to Mortimer and Scott (2003), who considered the authoritarian approach appropriate and to some extent necessary, I favored a dialogic approach. Dialogic pedagogy leads to humanizing teaching whereas authoritarian, dehumanizing teaching leads to authority-dependent and passive citizens (Frankenstein, 1983; Freire, 2013; Gutstein, 2006; McLaren & Kincheloe, 2007). As evident in the excerpt above, the students responded to my dialogic teaching by verifying the numeric accuracy of the calculations, identifying numeric patterns in the formula, and exchanging questions and answers. In short, the students actively participated in knowledge construction processes; their contributions and ideas were welcomed and valued, just as Nicholas and Bertram (2001) recommend.

Introducing EUP 5

I posted the paperwork for the project and demonstrated an example scenario for Edward's life after college. Each group had to agree on a study field for Edward. They then determined a possible amount of loans for tuition and made a payment plan by calculating the interest rate for five years. I handed out the paperwork for the task and allowed the students set their own groups depending on their choice of study field. Once the students began working on the project, their peer interactions intensified. Each group began by deciding on a study field and researching tuition in that field. They also factored in Edward's expenses over four years, such as textbooks, housing, food, and other personal items.

Thus each group determined the total amount of money Edward would need for the loan. Then they calculated the compound interest to determine an installment plan for paying it back in five years after graduation. Depending on Edward's job situation, they pictured different scenarios for social, economic, and cultural aspects of his future life. By the end of the period, three groups

STUDENT LOAN CRISES

had not completed their work. They asked if they could have lunch in my classroom while completing their project, which I agreed to.

I noticed that SLD was a burning issue for my students: their motivation was much higher than I anticipated. I observed intense peer discussions, exchanges of arguments, and collaboration on the calculations. As the students had to reach a consensus on multiple subjects, I had expected some conflicts or antagonistic arguments, but I was wrong. All the groups worked in harmony. The effort we, as a whole class, made throughout the past four projects was bearing fruit. We were making qualitative progress toward CME. Our class was now displaying qualities of a dialogic classroom and a community of collective learners. Since EUP 1, the whole class had attempted to achieve egalitarian collaboration, dialogic peer interactions, and horizontal student-teacher relationships. As we cumulatively applied what we learned from one project to the next, our learning snowballed.

Day 2

The second day began with online research. The students reviewed articles, commentaries, and statistical information about the SLD issue in the U.S. In this project, a different format for the whole-class discussion was trialed. First, each group shared their findings. Then, if the whole class agreed with their conclusion, a student volunteer typed them in a PowerPoint file on the laptop on my table while it was projected on the board. On this first occasion, Nicole volunteered. Once the file was completed, we ran through each slide from beginning to end; this led to whole-class discussion. Some of the information compiled about SLD is presented below:

– Over the last 20 years, the amount of SLD doubled.
– In 1994, less than 50% of the graduates had SLD; that number in 2015 is 70%.
– SLD has increased by 84% just since 2008, and in 2015 hit a grand total of approximately $1.2 trillion in the U.S.
– In 2014, four out of every 10 family men or women younger than 40 years old were paying off SLD.
– SLD is one of the root causes for new college graduates in the U.S. to delay marriage.
– From year 1970 to 2013, the average cost of college tuition increased by 275%.
– Since 2005, SLD has drastically increased, but salaries for young college graduates have decreased. It became difficult for young college graduates to get a decent-paying job. Many college graduates end up having a minimum-wage job.
– There are some countries in the world where public universities are tuition-free.

This activity seemed to expand the students' horizons and helped them situate SLD as their problem in the larger socioeconomic structure. None of the information or data about SLD surprised my students. They indicated that they knew someone either in their immediate family or relatives who was or is having trouble with SLD. However, the students did not want to believe that there are places in the world (about 40 countries) where college education is tuition-free or tuition costs a symbolic amount of money. I was surprised that my students were not aware of that information.

I observed that as much as the students appeared to be empowered by the discussion, they seemed worried and upset with the current SLD situation. While developing the PowerPoint slides, a spontaneous discussion in the class emerged. I wanted to bring up the consequences of market-driven education policies and discuss SLD in that context.

I did not, however, want to impose my viewpoint. Instead, I aimed to make some small openings in the colonized life-world of the students where they could be engaged in an open dialogue with their peers. When I asked whether education should be free for all or only for those who cannot afford it, without exception, all said that college should be free for all. For about 10 minutes, I observed the students posing questions, exchanging opinions, and answering each other's questions, all without interrupting each other. Unfortunately, I had to bring this discussion to an end and begin the journal entry session, as we were running out of time.

My ongoing reflections about this experience prompted the following question. In the context of critical pedagogy and CME, where should teachers draw the line between expressing their standpoint and making small openings for egalitarian dialogue? This question was raised by Shor (1993), one of leading figures in critical pedagogy. Based on his teaching experiences in a community college in the U.S., Shor (1993) elaborated:

> Dialogic teachers do not separate themselves from the dialogue. The teacher who relates economic power in society to the knowledge under inquiry in the classroom cannot impose her or his views on students but must present them inside a thematic discussion in language accessible to students, who have the freedom to question and disagree with the teacher's analysis. (p. 30)

Although I agreed with my students about the SLD issue, what I inferred from this dialogic experience aligns with Shor's (1993) point. As long as a discussion does not turn into a process of propaganda or manipulation of any kind, and

STUDENT LOAN CRISES

students can comfortably question the teacher's point of view, the teacher can participate in dialogue as an equal partner.

Students' Journals

This is the third time that Edward figured in our word problems. The students seemed to have much stronger empathy with him in this project than in previous ones. They imagined the next 10 years of Edward's life after college with debt. To provoke discussion, I posted a number of prompt questions on the board:

– How do you think paying back his student loan for 10 years will affect Edward's life?

Jacob thought that Edward's life would be restricted by his debt: "I picture Edward, struggling financially and living in an apartment with a roommate to save money after graduation as salaries are decreasing." Because the students determined Edward's field of study and estimated his tuition, each group's emphasis was slightly different. Selena mentioned her group's calculation and linked Edward's debt issue to her family's debt problem in the past:

> Edward's life doesn't seem so bright....According to our plan, he has to pay off $100,000. Whatever he does in his life, he has to think about his debt.... Being in debt is a horrible thing. When my father had an accident, he had to stay in bed for five months; we did not have any income and got into debt.

Darryl emphasized the connection between one's debt and getting married and starting a family: "It would be very difficult, if not impossible, to start a family in his strict financial situation." During the online research, the students compiled some statistical data, which confirmed Darryl's claim: SLD forces many college graduates to postpone their marriage plans.

Tibu voiced another common concern of students who are in debt: "Edward's life depends on many factors; his debt will definitely limit his choices, but if he gets a good paying job, he may pay off comfortably, but what if he doesn't get a job, that would be a nightmare." Jennifer had another term for what Tibu called a nightmare. She wrote, "Edward will most likely face the dark side of the American dream....And I hope he will not." Cindy provided a brief but vivid description of Edward's future life:

> Life after college for Edward may look like working long hours, maybe a second job, and restricted social life. He may not want to get married and have a child; he would not want his child to go through same problems....

> According to our project, Edward has to pay back approximately $20,000; he could buy a nice house with that money, but he will pay to loaners instead....It is not fair for anybody.

The students' entries consider Edward's next 10 years as stolen years. With the next question, I invited my students to share their experiences and opinions about socioeconomic consequences of SLD:

– How do you think debt in general affects people's life? Examples from your experiences?

In Akil's view, debt enslaves people: "Debt ruins people's lives....If you are in debt, you are kind of slave." Cindy shared her father's situation, which resulted from SLD: "My father works overtime to pay off my sister's loan debt and two of my cousins, whose parents are not fit to support." The students pointed out that SLD influences not only individual students, but also their families. As Akil articulated, debt can become the apparatus for social control and oppression.

With the following question, I invited the students to share their views about the neoliberal perspective on education:

– Do you consider education as a human right or a commodity?

From the neoliberal perspective, education, like any other item in market, can be bought and sold. In this view, education is considered a commodity rather than a human right (Chubb & Moe, 1990). However, the students seemed to disagree with this neoliberal tenet. I was expecting that a couple of students would disagree with an idea of tuition-free education. My expectations were wrong. All students argued that education should be a human right, and therefore free of tuition. If a citizen wants to study, there should be no financial barriers whatsoever to their educational journey.

For example, Nicole disagreed with the neoliberal ideology, arguing that education is a social investment. She wrote:

> If a student has determination and will, she should be able to go to college regardless of her money situation....When an individual is educated, the whole society benefits. So education should be a human right and social investment, not an individual commodity.

Darryl voiced a similar sentiment: "Everyone should be able to go to college regardless of their economic status. Education's cost should be shared by everyone because it is a necessary part of our society that affects everyone and everything." Cindy argued that education could not simply be reduced to

STUDENT LOAN CRISES

an item in the market: "Education can't be priced. It should be free for all....
We need to change our pledge...like liberty, justice, and free education for
all." Her comment seemed to imply that the pledge of allegiance[2] is empty
rhetoric: there could be no liberty and justice without education for all.

Leonardo was another who envisioned education as a basic right of citizens:

> Education should be a human right to give everyone a good career. But
> unfortunately, college education is something you pay a lot of money for
> in our country even though these days you have no guarantee that you're
> going to have a decent job after graduation.

Contrary to my expectations, students' arguments for free education were
emotional but nevertheless well-informed. Their comments challenged the
neoliberal stance that education is an commodity and thus each student
should pay to receive it.

As I observed the tension between neoliberal impositions and the students'
needs as human beings, I encouraged them to share their concerns and feelings:
– Does dealing with student loan problem in this project make you feel
   informed and empowered or instead make you worry about your college
   education?
The students articulated their thoughts and feelings about the SLD issue, as it
directly relates to their own life-world. For example, Nadia, like her classmates,
voiced her anger about the unfairness of college tuition in the U.S.: "Now I have
a better sense of my college education. But it is unfair and makes me angry. I
am not sure how, but I feel like this needs to be changed." Jacob indicated that
he felt empowered as well as informed about the problem; however, SLD made
him anxious, too:

> I usually worry about my future, and facts about SLD are startling. At the
> same time, I feel like I am aware of the problem. I don't think we can solve
> this problem individually. This is the first time I am experiencing a math
> problem that connects to the student loan problem.

Jacob's reflection also implies that the student loan issue is not critically
addressed in his education journey, as this project is the first for him.

Some students' reflections clearly indicated that being able to receive college
education in the U.S. is directly connected to one's socioeconomic situation.
Selena wrote, "The whole idea of SLD does worry me, because my family
does not have that much money. I am hoping to be qualified for a scholarship

maybe." Darryl voiced similar sentiments: "As much as I became informed, it made me worry about my college education and my future. My family is not that rich to pay my tuition if I go to a good college....Something needs to be done for high college tuitions."

Jennifer agreed with Darryl that something has to be done:

> I have mixed feelings; I feel like I am informed about a catastrophe of SLD and I am glad that I am aware of that. But it also scared me a lot...the way things are in this country is stifling our future....It is not fair....People in this country—I mean all of us—should do something about it.

She considers the socioeconomic and political system as the source of this unfair situation and indicates that citizens collectively need to take action to ensure that the system will no longer be able to destroy young people's futures.

The students' responses to this question revealed one of the challenges of critical pedagogy: How can hopelessness and pessimism be transformed into hope and love oriented toward ideas and actions for a more just and free world? If critical pedagogy fails to make this transition, it may do more harm than good, making students feel hopeless and helpless about their college education.

Therefore, the next question invited students to think of possible solutions to the problem:

– What do you think that can be done to remedy student loan debt issue?

Finally, the students shared their thoughts about possible solutions to the problem. Along with anger and frustration, they came up with suggestions aimed at solving it. Darryl envisioned a free college education and proposed a strategic step: "In my opinion, college tuitions should be significantly lowered and that would be a good step toward a free public education....Some countries offer free university education. Why not in the U.S.?" Through this project, the students learned that there are many countries in the world where university education is free: They were fascinated with this information.

The students also articulated the link between the job market and the SLD crisis. They wondered how college graduates could pay back their loans if there were not enough employment opportunities. Noah wrote, "Decreasing college tuitions and increasing minimum wage would be a good start to solve SLD crisis; also, you don't have to pay your loan until you get a job." Tibu put forward a similar perspective: "Creating more jobs, not charging any interest

STUDENT LOAN CRISES

rate for loans, and increasing the minimum wage would be a good start." The students argued that if more employment opportunities are created and the minimum wage is increased, it would eventually affect the jobs that require a college degree by increasing their salary.

Although most students considered SLD a crisis and came up with possible solutions, they did not to attempt to explain or question the root cause of the problem. Jeff was an exception. He pointed out that the first step should be to end corporate involvement in college education: "We need to find ways to remove big corporations, corruption, and greed from education....We need to speak out for better conditions and better opportunities for all." Jeff framed the issue as part of a larger struggle for a free and just society.

It was also surprising to me that some students pointed to excessive military spending.[3] Liam wrote: "I believe that tax money that is spent on military should be shared with funding for universities. This way it would lower the cost of going to university." Jennifer agreed, saying that taxpayers' money would be better used lowering SLD instead of killing people. She wrote, "Our army is dropping bombs overseas and says that they will bring democracy to these places, but would it not be better spending this money for universities rather than spending it on bombing people?"

Cindy pointed out that the system keeps people from publicly discussing problems such as SLD:

> To be honest, I don't know how the SLD crisis can be realistically tackled....
> These problems are not discussed in public; TVs are loaded with celebrity
> stories....American people are kept in dark. People need to be able to talk
> about issues like the SLD crisis, and something may come out of these
> discussions....But people think that they don't matter; for example, my
> parents don't vote, they think an election is completely meaningless.
> It doesn't change anything in their life.

Cindy articulated the liberating power of open dialogue. She implied that the system colonized the life-world and does not leave much space for people to be engaged in a free and open dialogue oriented toward possible solutions to real problems. Cindy's comment relates to Habermas's (1975, 1996) argument that neoliberalism maintains its hegemony through colonizing the life-worlds of people, shrinking the public sphere, and then imposing the system's imperatives top-down. This conclusion derived from students' comments about EUP 5 signifies the importance of creating small openings in the classroom where empowerment and liberation can emerge.

Concluding Discussion of EUP 5

In EUP 5, I integrated the standard curriculum (exponential modeling) into the SLD issue through an inquiry-based dialogical education without compromising rigor or academic standards. Freire and Faundez (1989) reasoned that "democracy and freedom are not denial of high academic standards" (pp. 33–34) and pointed out the misconception that academic rigor can be achieved only by authoritative teaching. A false connection is often made "between a democratic [dialogic] style and low academic standards... [and between] high academic standards and authoritarian [teaching] style" (pp. 33–34). My experience in this project is that dialogic teaching does not prevent high academic standards from being enacted. On the contrary, dialogic pedagogy makes the learning process more meaningful and empowering. For example, in EUP 5, as well as learning the interest rate formula, the students participated enthusiastically, which improved their collaboration and communication skills. Based on the students' comments and journals and my reflective journal, the students improved their content knowledge and seemed to develop critical mathematical literacy—one of main goals of mathematics education (Ernest, 2002b; Gutstein, 2006).

Instead of providing students with the interest rate formula and a couple of repetitive examples, I aimed to engage them in "problem-posing" in the sense of Shor (1993):

> In problem-posing, in teaching subject matter dialogically, academic material is integrated into students' life and thought. Students do not simply memorize academic information about [mathematics], but rather face problems from their lives and society through the special lens offered by an academic discipline. (p. 31)

SLD is a significant issue for students who wish to pursue university education. The students applied the interest rate formula as well as their mathematical literacy to envision social, cultural, and economic aspects of Edward's life and reflect on the SLD crisis in this context. During this project, I made small openings by raising questions about the SLD issue and encouraging students to do the same. In this sense, I critically distanced myself from mainstream mathematics instruction and textbooks. Instead, I constructed the project around three interrelated concepts oriented toward CME (Skovsmose, 1994): critical distance, critical competence, and critical engagement.

Through my teaching of exponential modeling, I concluded that introducing a new mathematical concept, theorem, or formula can be achieved through

STUDENT LOAN CRISES 105

dialogic pedagogy. This conclusion provides a precise answer to the question: "To what extent would CME agree with the idea that authoritative (or nondialogical) teaching is necessary when mathematics teachers teach a new subject?" The main point here is ethical and political. On ethical grounds, I prefer the dialogic approach because it is empowering and humanizing. The students actively contributed to creating calculation tables; they responded to each other's questions and comments and worked within multiple ZPDs. All of which is to say that although the interest rate formula already existed, it was not yet alive. Through dialogue, we collaboratively brought it to life by unpacking its internal logic and coherence. Instead of memorizing it, we developed the formula with understanding. Thus, *how* we learn mathematics is as important as *what* we learn.

Teaching the interest rate formula, I referred to a set of mathematical axioms, properties, and theorems. That is, I made a validity claim rather than a power claim, just as democracy itself must constantly refer to the Constitution for its validity check (Almeida, 2010). Because I want students to develop their participation skills, democratic values, and mathematical literacy en route to becoming critical citizens, I encouraged them to actively participate, check the accuracy of calculations and validity of conclusions, and take ownership of their learning. My approach is consistent with Ellis and Malloy (2007) and Hannaford (1998) who argue that mathematics education should promote critical literacy oriented toward critical citizenship.

From my observations and reflective journal, I noticed that ever since the beginning of the school year, we had been working to become a critical community of mathematics learners. We had made qualitative progress toward the goal of dialogic learning, developing egalitarian peer interactions and horizontal student-teacher relationships.

As noted earlier, the sharp increase in college tuition since 1980 in the U.S. is directly linked to the impact of neoliberal ideology on higher education. As Coco (2013) argued, neoliberal forces initiated "a shift away from education as a public good [to] education as an individual investment" (p. 566). These forces convinced "state and federal governments to disinvest in institutions of higher education" (p. 566). When universities received less public funding, tuition was raised drastically. Considering neoliberal policy the root cause of the SLD crisis, Coco (2013) argued that "this vision of education and the resulting education policies has led to tuition hikes, reductions in scholarships and grant funding for students, and an increase in students drawing on loans to pay tuition costs" (p. 566).

While these top-down neoliberal policies resulted in the SLD crisis, the students through this project developed bottom-up responses that radically challenged neoliberal educational tenets. The students' comments resonate with Coco's (2013) point that "many low- and middle-income students are either prevented from attending college, or they are required to assume an enormous debt loan, equivalent to a thirty-year home mortgage, to obtain a university education" (p. 601). My students challenged neoliberal ideology: They proposed steps oriented toward tuition-free education. They made tangible short-term suggestions to remedy the SLD crisis, such as creating more employment opportunities for college graduates and minimizing the interest rate for tuition.

The students' reflections indicated that they do not give their consent to neoliberal educational policies. According to Habermas (1975), this tension between the life-world and the system brings into question the very legitimacy of the system. He conceptualized this problem as the legitimation crisis. Habermas (1984c) concluded that this tension could only be solved by initiatives originating from the life-world; emerging bottom-up responses could transform the system. Creating small openings in colonized life-worlds—such as mathematics classrooms—is crucial to counter neoliberal hegemony and generate hope for a more just and free world.

Notes

1   Further information can be obtained at many web sources such as http://www.usatoday.com/story/money/personalfinance/2015/08/23/credit-dotcom-student-loan-crisis/32015421/; http://www.huffingtonpost.com/news/student-loan-crisis/; http://us.milliman.com/insight/insurance/The-student-loan-debt-crisis-in-perspective/

2   The pledge of allegiance is a ceremony held every morning in public schools in the U.S.: "I pledge allegiance to the Flag and to the Republic for which it stands: One Nation indivisible, with Liberty and Justice for all."

3   My observation confirms that the military budget is one of the taboos in the U.S.; it takes intellectual capacity and civic courage to speak in public about overinflated military spending.

CHAPTER 7

Critical Mathematics Education: A Bottom-up Response to Neoliberal Hegemony

In this chapter, I address the question, "What are the potentials and limitations of critical mathematics education in terms of classroom teaching in the neo-liberal era?" I present a conceptual analysis of the end-of-unit projects (EUPS). From a holistic perspective, I reconceptualize CME as a counterhegemonic practice of teaching and learning mathematics. Analysis of the findings from the EUPS suggests that three concepts are especially important: (a) dialogue as pedagogy, (b) collaborative learning, and (c) inquiry-based education. In the education literature, these terms typically reference mainstream educational perspectives. Therefore, it is necessary to define the distinctive elements of dialogic pedagogy, collaborative learning, and inquiry-based education in order to elucidate their scope in a CME approach to mathematics in the high school setting.

My students and I began practicing democracy in the form of dialogue in a mathematics classroom that had been previously colonized by top-down neo-liberal educational policies. In this analysis, I therefore recognize neoliberalism as an ideological and pedagogical construct in order to understand the limitations of CME. The chapter illustrates certain challenges and obstacles that math teachers may experience when they practice CME.

Critical Mathematics Education and Dialogic Pedagogy

Dialogic pedagogy played a defining role in the process of meaning making and knowledge creation in this study. For example, during EUP 1, the students in their reflections and evaluations identified non-egalitarian peer relationships as a barrier to establishing dialogue. As a response to this finding, the class made collaborative attempts to overcome this barrier through cycles of reflective actions. As each EUP unfolded in response to what had been previously experienced, it seemed that a more dialogic classroom was developing, as evidenced in student reflections and in my reflective journal. Across all five EUPs, this transformation was especially noticeable: the students became more proficient in exchanging ideas, deconstructing and reconstructing their values, attitudes, and ideas. This finding mirrors Frankenstein's (2010) teaching

© KONINKLIJKE BRILL NV, LEIDEN, 2019 | DOI:10.1163/9789004390232_007

experience in a U.S. community college, where she used statistical data to unpack the oppressive and discriminatory structure of society.

Approaches to dialogue in education can be divided into two main ideological perspectives. Nicholas and Bertram (2001) explain:

> The contemporary vision of dialogue as a pedagogy that is egalitarian, open-ended, politically empowering, and based on the co-construction of knowledge, reflects only certain strands of its history. Contrasting accounts see dialogue as a way of leading others to pre-formed conclusions; or as a way for a master teacher to guide the explorations of a novice; or as a set of ground rules and procedures for debating the merits of alternative views. (p. 1102)

This statement divides dialogical approaches into two main lines of thought. Dialogue as communicative rationality (the contemporary vision) and dialogue as instrumental rationality (the contrasting account). The theoretical underpinnings of my research, complementary with Freire's and Habermas's ideas, are inspired by the former strand, which embraces dialogue as communicative rationality. From this point of departure, dialogue as pedagogy is endorsed because it democratically promotes horizontal teacher-student relationships; it opens up possibilities by promoting critical and creative thinking; it is liberating by helping students to develop communicative competency and critical consciousness; it is a collaborative process through which students can learn with and from each other. The essence of this approach is that it seeks ways of establishing non-dominating communications and relationships in an educational context oriented toward critical literacy and humanizing education. Dialogue, hence, is not only an effective way of learning, but also an end in itself. As such, Freire (2000) considers dialogue to be an existential necessity.

The contrasting account sees dialogue through a lens of technical (instrumental) rationality, considering it to be an effective way to transmit prepackaged knowledge to students. Habermas (1984) considers this form of communication, in which a preset agenda is imposed on learners, as distorted; similarly, Freire defines it as "banking" education—an oppressive pedagogy in which an anti-dialogical process disseminates dehumanizing education (Darder, 2002; Freire, 2013; McLaren & Kincheloe, 2007; Shor, 1987). Therefore, pedagogy based on dialogue in this sense may promote functional literacy, adaptation, and domestication.

Through small-group discussions across cycles of actions, the students became increasingly willing to push their comfort zones and revise misconceptions they might have had. This adjustment seemed to improve their conceptual understanding of mathematics. For example, during EUP 1, many students said that they identified some misconceptions and improved their understanding of linear functions and equations. Through whole-class discussions and written reflections, the students began to relate their life in the classroom to the system—namely, to externally imposed standardized assessment as the measure of learning. In other words, as the students engaged with mathematics content learning, each EUP and the dialogue that constituted it connected students' lives in the classroom to a larger social, political, and economic system, thereby promoting critical mathematics literacy. Dialogue inspired by an "ideal speech situation" Habermas (1990, 2005) helped create the necessary freedom for students to construct knowledge on their own volition, without it being imposed by a more powerful agent.

The theory of communicative rationality as framed by Habermas (1984, 1987) is complementary to dialogic and humanizing education as outlined by Freire (1998, 2000, 2013). Across EUPS, dialogue as experienced by students seemed to promote critical literacy and transformative action as well as functional literacy. Therefore, dialogue—as an approach to teaching and learning mathematics—was here based on *communicative* rationality, which is fundamentally different from the *instrumental* rationality approaches that students may have experienced previously (Nicholas & Bertram, 2001). In this study, dialogue was not used as an instrument to transmit pre-formed knowledge or improve standardized test results.

Results showed the specific ways in which dialogue in the CME approach constituted a distinctive and new form (for myself and my students) in its relation to learning, curriculum, peer interactions, students-teacher relations, and overall classroom ambiance.

Open-Ended Projects as Mathematics Lessons

An important element of this study's dialogic pedagogy was open-ended word problems that I designed. In the traditional approach to teaching mathematics, skill-drill types of repetitive exercises are favored. Solutions to these questions require procedural knowledge only; they lead to one-dimensional, single correct answers. Thus, such approaches foster rote memorization. Thanks to top-down neoliberal educational policies, this "exercise" paradigm (Skovsmose, 1994, 2011; Skovsmose & Alrø, 2004) has become the driving force of most mathematics classes (Hursh, 2007b; Gutstein, 2006). In an era of market-driven policies and curriculums, teachers are urged to focus on exercises that will most likely be assessed on standardized tests (Leistyna, 2007; Schneider, 2015). Consequently, student-teacher communication that emerges in this context draws on instrumental rationality and implies strategic action. Habermas (1984, 1987) defines such communication as "distorted"; Freire (2000) calls it "anti-dialogical."

In contrast to the traditional approach to mathematics education, I designed the EUPs as open-ended projects. This allows multiple perspectives to interact; it creates a learning ambiance conducive to dialogue. For example, in EUP 1, students produced their own multiple-choice questions to show their conceptual understanding, procedural knowledge, and numerical fluency with linear functions and equations. The project did not require the students to follow a prescribed direction that would lead to a single correct answer. Instead, it was open-ended, allowing each group to collaboratively negotiate possible choices—both correct and incorrect ones—and agree on one set of choices. My choice of inquiry design, as a teacher, was thus not based on an instrumental rationality but rather communicative rationality.

Although direct teaching (lecturing) in mathematics has been suggested to improve students' learning of mechanical aspects of mathematics and enhance their functional literacy, my experience as a mathematics teacher suggested it was more appropriate to teach linear functions through an inquiry-based dialogic lesson. The dialogue, coupled with the open-ended task, appeared to be a liberating dynamic for the process of knowledge construction. With an inquiry-based approach, I aimed to open a dialogic space in my classroom where the students would be able to negotiate with their peers and develop their own solutions.

Inquiry-based processes seemed to enhance the quality of learning in this class. Each EUP involved an inquiry that allowed students to apply and improve a range of algebraic concepts, as well as connect their learning to a larger sociopolitical structure. This was evident in whole-class discussions and in students' reflective journals. In order to connect content learning to social, economic, and political issues, a shift from a problem-solving to a problem-*posing* approach is required (Gutstein, 2006; Skovsmose & Alrø, 2004). In EUP 1, for example, an investigative approach provided students with opportunities to unpack standards-based assessment and evaluate them in terms of the needs of individual students.

The students' journals and classroom discussions confirmed that peer and student-teacher communication was a key to knowledge construction. For Freire, knowledge is socially constructed through "subject-subject dialogue"; for Habermas, it is built through "intersubjective consensus within discursive communities" (cited in Morrow & Torres, 2002, p. 54). The students' reflections on group work indicated that peer interactions enabled them to learn with and from each other. For example, in EUP 2, the process of developing the multipart function encouraged students to exchange thoughts with their peers, and listen to, and build upon, each other's ideas. I conclude that open-ended word problems create dialogic spaces for students to challenge the hegemonic dynamics that affect their life-world in and out of the classroom. In short, CME can promote critical consciousness (Freire, 2000) and communicative competency (Habermas, 1987).

Dialogue within Inquiry

Results revealed four fundamental elements of an inquiry-based approach to mathematics education. First, themes of inquiry must be derived from students' life-world in order to motivate them and increase the quality of their communication and participation. I noted that the students readily joined discussions about social and political matters embedded in EUPS when the topic related to their social circumstances. Second, the inquiry process must welcome ideas that may be unpopular within the mainstream curriculum. Third, the inquiry process must be facilitated by the teacher to enable students' critical engagement. Fourth, the process must include a self-evaluation (reflection) as an alternative to assessment in the mainstream system. Traditional education has its own measurement system—normally a letter grade or percentage. Therefore, an inquiry-based approach within CME should also include a critical reflection and an evaluation process to ensure that CME is sustainable as an educational practice, albeit a subversive one.

In the process of conducting the action research cycles, I realized that the whole-class discussion could function as a process of self-evaluation for students' learning as part of CME practice. This finding is in line with Rogers's (1995) notion of person-centered education. The whole-class discussions at the end of each EUP gradually became critical self-evaluations, in which plans for the next project were initiated within the continuous cycle of planning, acting, and reflecting. Just as whole-class discussions became a process of collective evaluation, students' reflective journals facilitated an individual evaluation process. These collective and individual reflection processes dialectally informed each other throughout all five projects.

Power Relations and Dialogue

During EUP 1, students identified unequal peer relationships as an obstacle to establishing productive dialogue in small-group work. In response, they, with my encouragement and scaffolding, worked on establishing non-dominating peer interactions. The measure of success of these endeavors is that in subsequent EUPS, students reflected on *equal* power relations in group work. Thus, the data show that the nature of communication among students became a distinctive element of dialogue in CME.

In the traditional approach, communication between students and teacher occurs within an asymmetrical power relationship. This power relation is considered normal: Teachers possesses authority and knowledge; they transmit it to students, who passively receive the transmitted knowledge (Rogers, 1995). Notwithstanding widespread acceptance, this approach creates a toxic, anti-dialogic atmosphere in the classroom (Shor, 1987).

My research, however, strongly confirms Freire's (2000, 2013) and Rogers's (1995) conclusions; namely, that teachers can turn a classroom into a supportive and liberating environment. Throughout the study, I aimed to be a facilitator who understands students empathetically, who is an equal partner in the collaborative learning community, and who establishes horizontal relationships with the students. As I became a better facilitator, the classroom ambiance became friendlier and more positive. In their journals, some students said that my facilitative attitude had a positive impact on peer interactions. But whatever the cause, students deliberately and enthusiastically worked on developing non-dominating peer relationships throughout the project. They felt themselves to be an integral part of the classroom and related to their classmates with empathy.

As I shared my power with students and fostered a facilitative learning environment, the typical discipline problems and classroom management issues that I've experienced over the last seventeen years seemed to mostly dissolve. This observation led to me to conclude that how students behave is connected with whether or not a classroom is dialogic.

Resonating with Freire's notion of humanizing education, Rogers and Freiberg (1994) considered a teacher to be a facilitator who is a key to achieving person-centered instruction. According to Rogers and Freiberg, "There is no resemblance between the traditional [authoritative] function of teaching and the function of the facilitator of learning [dialogic]" (p. 170). A teacher's facilitative stance is not based on a power claim. Rather, teachers are "catalyzers, facilitators, energizers; they give students freedom and life and the opportunity to learn. Most importantly, they are co-learners with students" (p. 167). Similarly, Freire (2013) recommended the empathetic approach and horizontal student-teacher relationships to achieve liberating communication. According to him, dialogue is the only real communication as "it is nourished by love, humility, hope, and mutual trust" (p. 42). Freire (2000, 2013) expressed the centrality of trust in humanity and indicated its transformative potential.

Rogers and Freiberg (1994) elaborated the teacher's role as facilitator and its impact on peer relationships:

> When teachers are empathically understanding, their students tend to like each other better. In an understanding classroom climate, every student tends to feel liked by all the others and has more positive attitudes

toward self and school. This ripple aspect of the teacher's attitude is proactive and significant. (p. 161)

Findings from my classroom support Rogers and Freiberg's (1994) conclusion. For example, in EUP 1, the students identified dominating attitudes as a barrier to productive and enjoyable group work; as a response, they enthusiastically worked on developing egalitarian peer relations. Initially, as recorded in my reflective journal, I considered it a coincidence that the students were working well together, listening to each other respectfully, and being kind to each other. However, I gradually realized that my empathetic approach—as opposed to taking an evaluative stance—seemed to inspire students to embrace non-dominating peer interactions in subsequent EUPs. As Freire (2000) pointed out, mutual trust is key to establishing a facilitative attitude and classroom ambiance. After five cycles of action research, I concluded that the preliminary reconnaissance stage played a significant role in establishing grounds for mutual respect and trust.

In line with Gutstein's (2006) and Frankenstein's (1994) suggestion that CME should serve critical mathematical literacy, students connected their mathematics knowledge and life in the classroom to the social, political, and economic system to interrogate unjust and oppressive aspects of the system. Findings illustrate the ways in which these discussions created small openings to counter neoliberal hegemony. Therefore, the discussions inevitably involved ideological and political views.

Contrary to traditional classrooms, where teachers are not supposed to express their views, I openly shared my point of view. In EUP 1, for example, I shared my ideas about standardized assessment; I shared my views of socioeconomic inequality in EUP 4 and tuition-free education in EUP 5. As a facilitator and equal partner in the knowledge construction process, I provided them with opportunities to question and criticize my standpoint. Although it was not antagonistic, I observed that some students opposed their classmates' views as much as mine, which contributed to the process of creating a dialogic classroom. My experience is consistent with Freire's (2000) conclusion that critical teachers can openly share their views as long as they can establish and sustain horizontal relationships with the students through which they can freely disagree with the teacher's view.

Neoliberal educational changes have resulted in the implementation of a scripted curricular sequence (Hyslop-Margison & Thayer, 2009; McNeil, 2009; Schneider, 2015). I incorporated the script into each EUP. For example, students needed to be familiar with basic properties of exponential expressions, equations, and functions in order to pass the standardized test. Therefore,

EUP 5 required students to understand and use exponential expressions. The CME literature is silent about whether mechanical aspects of mathematics can be taught through a dialogic approach.

Prior to this study, as a high school mathematics teacher, I had often used authoritative[1] approaches in my teaching practices. Such approaches urge students to memorize and accept what the teacher says, with or without understanding. Over time, I began to notice that the authoritarian approach generated frustration, anxiety, and boredom among my students. Furthermore, it seemed very difficult to learn mathematics through memorization. In my ongoing professional development and reading, I came to the understanding that authoritative teaching was nurturing neither an enjoyable nor a humanizing learning experience for students. When I discovered the critical pedagogy literature and CME on my teaching journey, I began to practice dialogic teaching of mathematics.

In the present research, I applied dialogic teaching even when I taught mechanical aspects of mathematics. As there was no classroom-based CME literature to use as a guide, I had to develop my own dialogic teaching practices from the ground up. For example, in EUP 5, students needed to have a basic understanding of exponential properties to work on Edward's college application. I facilitated a process of collective construction of the interest formula. I demonstrated numeric patterns in order to formulate a generalization. For each step, I posed prompt questions to the whole class, which students voluntarily responded to. However, this experience turned into much more than typical student-teacher talk: what began as a single student-teacher exchange evolved into a whole-class dialogue. Students actively listened to and responded to each other with their own arguments and ideas; they ultimately took control and ownership of their collaborative knowledge construction process.

In this facilitative teaching, I did not continue to the next step until all students indicated that they understood the concept under discussion. While I was demonstrating numeric patterns, the students made all the calculations and verifications as active participants of this knowledge construction process. When I answered students' questions and cleared up their confusions, I justified my conclusions by means of algebraic properties and other mathematical theorems—not by asserting my authority as classroom teacher.

The lesson exceeded my expectations. The students were engaged. They actively participated in the process of reinventing the interest formula. They were as excited as if we had developed a new formula. My dialogic approach seemed to create a different classroom atmosphere. I noted that it would be difficult, probably impossible, to achieve this result if we were not a community. After each EUP, being a community collectively learning mathematics nurtured this authentic ambiance. It is evident that being a community enabled us to co-construct the formula in EUP 5. It is difficult to see how a traditional classroom—where students are positioned as passive receivers of knowledge, ruled by the authority (Rogers, 1995)—could have done the same.

The Mathematics Classroom as a Community

Taken together, the findings from the EUPs indicate that teaching a mathematical concept in dialogic form requires the classroom to be an egalitarian community of learners (Kennedy, 2009; Murphy & Fleming, 2010). My reflection suggested a dialectic connection between becoming a community of learners and quality of dialogue. Five elements of this connection emerged.

First, the teacher's role as facilitator is to establish and sustain a horizontal relationship with students. I noted that my facilitative attitude was one of the tipping points that reshaped the entire classroom ambiance. Acting as a co-constructer of knowledge rather than the ultimate source of knowledge and authority, I opened up space in the class, which allowed the students to become more agentic actors. As student journals indicated, this space encouraged students' voluntary participation. It enabled them to take ownership of their learning and develop responsibility, self-discipline, and egalitarian peer interactions.

Second, structuring learning as a collaborative (non-competitive) activity created a friendly environment where students did not see their peers as rivals and, therefore, were able to empathetically understand them. In this friendly environment, student-teacher conversations were transformed into whole-class dialogue. I observed that as my relationship with the students became more horizontal, peer interactions in the class became more empathetic and positive—a transformation that resonates with ideas of Rogers and Freiberg (1994) and Rogers (1995), who found that a teacher's facilitative attitude promotes more dialogic and empathetic peer relationships.

Third, there is a close connection between becoming a dialogic classroom and establishing mutual respect and responsibility. My facilitative attitude generated a liberating atmosphere through which the students developed self-discipline. Students took responsibility for and ownership of their own learning; thus, learning activities became a self-regulated process. In traditional, authoritarian education, students are regarded as strangers who need to be controlled (Kohn, 1999, 2006; Rogers, 1995). Not so in the dialogic classroom. As my students and I established mutual respect, empathy, and trust, I could rely on the students' responses as to whether to continue to the next step in learning activities. Students, meanwhile, exchanged ideas and comments without hesitation or fear due to this friendly environment.

Fourth, the students' being a community, in which learning is a collaborative and open-ended process, changed their perceptions about what it means to be confused or make mistakes. At the reconnaissance stage, we as a whole class agreed that making mistakes and being confused are a normal part of learning and not subject to criticism in our class.[2] As a result, students gradually felt free to share their ideas during group work and whole-class discussions.

Fifth, my efforts to establish horizontal student-teacher relationships acted as a catalyst for students to create egalitarian, non-dominating peer interactions of their own. In EUP 1, they agreed on equal peer relationships as a regulative norm, and continued acting upon this principle for each subsequent EUP.

Democracy and Dialogue

Dialogic pedagogy has implications for the connection between mathematics education and democracy. Almeida (2010) argued that teaching mathematical proofs is a domain where dialogic teaching can be materialized in a form of democracy. A proof refers to mathematical axioms, theorems, and properties in order to justify each step in the process. Similarly, in democratic societies, tensions and problems between citizens and authorities are resolved based on the country's Constitution. Viewing the mathematics classroom as a micro society, a teacher can structure the proof process on the "constitutional" basis of axioms, theorems, and properties, instead of on the teacher's authoritarian power. The National Council of Teachers of Mathematics (2000) also suggested that mathematics teachers construct teaching through the mathematical verification process—not their authoritarian power.

In EUP 5, in response to the students' question, we constructed a proof of zero power property: $X^0 = 1$. Inspired by Almeida's (2010) suggestion, I broke the proof process into several steps and negotiated each step with the students by justifying it through algebraic properties and logical verification. I ensured that every student clearly understood before continuing. My reflection on EUP 5 supports Almeida's claim. The dialogic approach enhanced the existing democratic culture in the class because the validity of the claim was based on not my authority as teacher, but on the mathematical logic, properties, and evidence of the proof itself.

According to Alexander (2006, 2015), teaching is dialogic if it is collective, reciprocal, supportive, cumulative, and purposeful. EUPs were based on collaborative learning and emphasized the value of collectivity. The students were equal partners in the decision-making process. It was evident that we established horizontal student-teacher relations and non-dominating and reciprocal peer interactions. The students expressed their opinions, ideas, and

A BOTTOM-UP RESPONSE TO NEOLIBERAL HEGEMONY

suggestions during group work and whole-class discussion without internal constraints, such as being afraid of making a wrong answer. We established a facilitative and supportive environment. Our projects were cumulative and consistent. At the end of each EUP, we reflected on our experiences and jointly planned the next project. The projects included a specific focus on mathematics content and its socioeconomic and sociopolitical context. Considering all the EUPs together, the findings confirm that the classroom was a dialogic sphere.

I conclude that Alexander's (2006, 2015) framework for elements of a dialogic classroom resonates with CME's vision. CME promotes critical literacy, aiming to provide students with opportunities to relate their learning experience to understanding and challenging oppressive and undemocratic elements in their life-world (Skovsmose, 2011). Taking Skovsmose as a point of departure, I derived the themes of all EUPs from the students' life-world, which enabled them to contextualize their learning experience within a larger system.

Habermas (1984, 1987) provided a comprehensive and interdisciplinary epistemological and ontological framework that elucidates conditions under which non-dominating communication could be materialized. He conceptualized this notion as the "ideal speech situation" that enables a consensus reached by noncoercive acceptance of better argument. Kennedy (2009) argued that the realization of an ideal speech situation requires a community of inquiry, in which inquiry proceeds through dialogue. The EUP sessions in my study confirm Kennedy's point.

The reflections—specifically, students' reflections—indicated that empathetic understanding, both between student and teacher and among peers, is a necessary element to fulfill the conditions for "ideal speech." In all EUPs, I observed that empathic understanding played as significant role in developing non-dominating peer communication. Darryl's journal entry in EUP 2 is relevant here:

> As soon as I looked at Edward's job offer, I figured out coordinates of the intercept point....You know, it was 10 hours that adds up to the same weekly salary when you plug them in both functions....But I did not say anything at first. I waited for my buddies to take time and think it through. Otherwise, I would spoil the moment. Instead I did the kind of loud thinking, just like Mr. Bülent helps us to ask questions to figure out what we need to solve the question.

Darryl indicates that he acted with a sense of responsibility and empathy that comes from being a community. From this incident, I concluded that if a group were not a community that promotes empathetic understanding, peer interactions could easily produce power relations and hierarchies. Darryl's empathetic approach can be applied to different situations. For example, a student whose mathematics knowledge and skills are above those of the others in a group may tend to dominate others; a student who is a native English speaker may consider non-native speakers less competent. Under these circumstances, reaching a consensus through better argument would fail to meet the conditions for an ideal speech situation.

Having facilitated five EUPs, I conclude that a dialogic approach to teaching and learning mathematics in a high school context, drawing on communicative rationality, can be achieved. It happens when a community of learners is formed, in which the teacher is a facilitator, where peer relations are egalitarian and empathetic, and where learning is collaborative and driven by inquiry. Dialogic pedagogy helped students develop both functional literacy and critical mathematics literacy, as well as communicative competency oriented toward critical citizenship. Dialogue as pedagogy in CME enables students to disagree with their peers and teacher; it must consciously and systematically reject imposition and propaganda of any kind when life-worlds and the system are connected.

Critical Mathematics Education and Collaborative Learning

Collaborative learning in the current study was organized through small-group work and whole-class discussion. In each EUP, students reflected on their experiences of group work to improve the overall quality of collaborative learning; they also acted upon their reflections to transform the classroom into a dialogic one. EUPs gave students the chance to negotiate egalitarian collaboration as opposed to competitive learning in learning mathematics. In EUPs 3 and 4, the whole class discussed social and political implications of collaboration and competition, which seemed to help them to make a conscious effort to turn their group work into a more equal collaboration over time. Their reflective journals provided evidence that across all EUPs, they made genuine efforts to improve the quality of peer collaboration.

My observations and students' reflections confirmed that the quality of collaboration in small-group work did indeed improve. From this finding, I conclude that the collaborative approach to learning of mathematics is a continuous pedagogy of praxis. The quality of collaborative learning can always be improved. As Pine (2009) and Groundwater-Smith et al. (2003) indicated, it requires time and systematic effort to establish a classroom environment that fosters collaborative learning.

At the reconnaissance stage, I asked the students whether they would prefer individual, collaborative, or competitive learning. All students favored collaboration. Students' journal entries across EUPs indicated that they collectively and consciously developed a culture of respect, inclusion, empathy, and listening to each other's ideas. My field notes also revealed that EUPs enabled students to improve their content-based communication. As these interactions turned into dialogues oriented toward specific educational objectives, it was evident that students' learning and empowerment were emergent throughout the process. The students wrote in their journals that EUPs enabled them to correct some mathematical misconceptions, and made them see themselves as an integral part of a community of collaborative learning.

Skovsmose and Alrø (2004) argued that dialogue and collaboration cannot be imposed on students—any meaningful participation in collaborative learning has to be a voluntary action. Findings revealed that it was important for students to reflect on group work experiences in the past to revise their attitudes about collaboration. For example, in EUP 1, students identified power relations as the main obstacle to productive and enjoyable collaboration. Students described "group" experiences in which everyone worked on a different task, or everyone worked on their own while sitting around the same table.

Students also described graded projects in which one member did all the work but everyone received credit. When I asked students their opinions about possible solutions to unequal power relations, they suggested that ensuring equal power for each member would overcome the problem. I had intended to make this very proposal, but the students were faster at becoming agentic participants. After EUP 1, establishing egalitarian collaboration became one of our shared objectives.

With respect to barriers to the collaborative learning of mathematics, Horn (2014) concluded from her ethnographic study that students' negative experiences with previous group work—due to unequal power relations—might be a serious obstacle. Horn also implied that neoliberal ideology is contrary to the collaborative approach. If students are allowed to derive power from their social or cultural status to dominate their peers, truly collaborative learning will not succeed. In that case, "Over time, this system may reinforce negative ideas [students] have about themselves as mathematics learners, because they may conclude that their ideas are not valuable" (Horn, 2014, p. 21). Findings from the current study underscore that students' agreement on egalitarian interaction is key in establishing collaborative learning.

Findings also supported the claim that collaboration cannot be imposed (Skovsmose & Alrø, 2004). Collaboration involves sharing, helping others, and non-dominating interactions—in short, learning with and from each other. But what if students do not consider these values to be virtues? To be proactive and overcome a potential barrier, I held whole-class discussions to provide the students with an opportunity to negotiate educational and sociocultural implications of egalitarian collaboration to enable them to make a conscious decision about their voluntary participation in small-group work. For example, in EUP 3, we discussed the productivity of collaboration versus competition in the context of the history of mathematics. The students agreed that collaboration has been much more beneficial and created more prosperity for humanity. In EUP 4, we discussed socioeconomic inequality, the caring community, and the virtue of helping those in need. Based on students' comments, these whole-class discussions seemed to expand their horizons about collaboration not just in the classroom but in society generally.

The literature shows that collaborative learning is difficult to realize within traditional approaches (Alexander, 2006; Kohn, 1992; Rogers & Freiberg, 1994; Smith & McGregor). Horn (2014) noted that students may bring existing power relations to group work; teachers should intervene if they hope to set up truly collaborative learning. Unfortunately, she did not provide any practical suggestions for such intervention.

The CME literature offers a very limited number of classroom-based studies about the dynamics of collaborative learning oriented to promote critical citizenship. Against this background, my research makes an original contribution. In order to prevent unequal power relations in group work, a cultural and ideological confrontation with neoliberal pedagogy is necessary. If the classroom is to be transformed into a community of learning, students need to accept, through whole-class discussions, the need for egalitarian interactions. As evidenced in my reflective journal, my intervention to establish collaboration drew on such virtues as mutual respect, sincerity, and transparency. This finding is consistent with Freire (2000), who noted that these kinds of interventions must not involve manipulation, imposition, or propaganda of any kind.

Collaborative Learning in EUPs

Over time, students embraced egalitarian collaboration as humanizing and productive, which helped transform our classroom into a community of mathematics learners. As EUPs opened a space in the class to talk about sociopolitical issues such as inequality, poverty, unemployment, and homelessness—taking Edward's story as a point of departure—the students strongly countered neoliberal ideology.

For each EUP, I facilitated whole-class discussions in which we negotiated the implications of collaborative (versus competitive) learning, caring for others, and developing non-dominating peer communication. In EUP 1, students agreed that egalitarian interaction is necessary to establish effective and empowering collaborative learning processes. In EUP 3, students considered mathematics as an outcome of the collaborative work of humankind throughout history.

Although we did not practice competitive learning here, the students criticized competitive learning based on their experiences in other classes. The students' reflections revealed that they participated more actively and learned with greater understanding because our class was collaborative. Students did not consider their classmates to be obstacles to their success. On the contrary, students said that they developed empathy, transferable skills, and knowledge through collaborative learning. This finding is consistent with Alexander (2015), Kohn (1992) and Rogers and Freiberg (1994), who argued that collaborative learning helps students both academically and socially.

EUPs were not evaluated: students were not rewarded with a grade in the traditional sense. Thus, students' motivation to help each other did not arise from extrinsic rewards. The source of motivation was the learning process itself. This supports Kohn's (1992) argument that extrinsic rewards are unnecessary to motivate students—collaborative learning is collaborative learning is its own reward

Neoliberal Pedagogy in Collaborative and Competitive learning

Cultivation of a competitive culture is one of the main objectives of mainstream American education. However, excellence in learning does not depend on competition. My professional experience confirms Kohn's (1992) point that any educational success that can be achieved by competitive learning can also be attained by collaborative learning. Students in American schools are introduced to the idea of competitive learning based on an ontological assumption that human beings are competitive by nature (Kohn, 1992). That assumption has ideological implications that downplay the idea of collectivity. Therefore, collaborative learning as practice of critical pedagogy necessitates a culture war and must be counterhegemonic. The practice may differ depending on classroom context. However, I should emphasize some points that contribute to a praxis of collaborative education within critical pedagogy.

First, the culture of collaborative learning cannot be achieved overnight: it takes time and students' active participation. Any educational effort for collaborative learning will fail if students doubt the value of collectivity and they just work with their classmates because they're required to do so. Therefore, a whole-class discussion to review practical, social, and philosophical implications of collaborative and competitive learning can be a good start toward cultivating a culture of collaborative learning. Through this discussion, as we experienced in EUP 3, students can deconstruct and reconstruct their ideas about value of collectivity compared to competition.

Second, collaborative learning as a practice of critical pedagogy must be structured through egalitarian peer interactions. Therefore, the discussion should also entail identifying whatever seems to be an obstacle to developing egalitarian peer collaboration. For example, students should be introduced to the notion that we need to find unity in diversity, as Freire (2000) points out: a perspective that considers our differences—ethnic, racial, religious, etc.—as something that enriches our lives rather than separates us from each other. In doing so, students with diverse backgrounds can find ways to develop non-dominating peer interactions and learn from and with each other.

Third, for business-related pragmatic reasons, neoliberal pedagogy somewhat promotes collaborative learning, too. Corporations would also like students (future employees) have skills to work with others to achieve business objectives. However, in neoliberal pedagogy, students complete externally assigned assignments without questioning the task itself. Therefore, it is important to distinguish collaborative learning in critical pedagogy and collaborative learning in neoliberal pedagogy. From the perspective of critical pedagogy, students doing collaborative learning plan their own work, set their own learning agendas, and question learning objectives; each student in a small group contributes to the joint work but preserves their autonomy. In other words, collaborative learning in critical pedagogy is not a process of "collective compliance" where individual students are lost in group work. Critical educators should be aware of this delicate and dialectic balance between collectivity and individuality.

Furthermore, even while urging students to work "collaboratively," assessment in traditional education is competitive and individual. These types of contradictions are natural consequences of neoliberal teaching practices. In contrast, the critical pedagogy teacher must create space in the colonized life-world of students, openly negotiating these matters with them to form a culture of collectivity in the classroom, and thereby countering neoliberal hegemony.

Inclusion, Facilitation, and Zones of Proximal Development

In EUPS 1 and 2, we discussed potential causes of exclusion in group work and together established classroom norms. The students identified three reasons for not participating:

- A self-assigned leader may dominate, resulting in less participation for everyone else;

A BOTTOM-UP RESPONSE TO NEOLIBERAL HEGEMONY

- Students' cultural or socioeconomic background might be a reason for being shy or withdrawn;
- Students who perceive their mathematics competence as inferior may choose not to participate.

The students' recommendation for equal power for each member dissolved the first issue effectively. In all EUPs, students ensured that a dominant attitude was unwelcomed in group work. For the second and third concerns, for each EUP, I facilitated whole-class discussions to clarify that each student in our class is a unique and equal member of our learning community regardless of gender, race, ethnic group, economic background, or mathematics ability. As the classroom gradually transformed into a community of learners, this point became a routine part of our practice rather than an intellectual abstraction. The whole-class discussions went a long way toward making group work inclusive and egalitarian.

As we worked to establish democratic peer relationships and positive interdependence, collaborative learning in small-group work became more inclusive. All students were welcomed, regardless of their mathematics ability or socioeconomic and cultural background. I noticed that students whose favorite subject was not necessarily mathematics were still able to actively participate in group work. The inclusive nature of the process seemed to enable these students to learn with each other. Thus, collaborative learning processes helped transform the classroom into a community of learners. This finding is in accord with Kohn (1992), who suggested that collaborative learning is a necessary element of inclusive group work.

As for my role, whenever I answered questions or provided scaffolding, I attempted to act as another member of the group instead of an authority figure. I answered questions with mathematical and logical justifications, but left space for them to disagree with or question my approach, logic, or mathematical argument. The students noted in journal entries that my facilitative attitude motivated them to learn with and from each other without trying to dominate or compete with one another. This finding further supports the ideas of scholars such as Groundwater-Smith et al. (2003) and Rogers and Freiberg (1994), who suggested that a teacher's facilitative attitude plays a significant role in transforming a traditional classroom into a collaborative community.

Another way of looking at classroom interactions is to use Vygotsky's (1978) notion of the zone of proximal development (ZPD). In critical pedagogy, an asymmetrical power relation is considered the main obstacle to dialogue, freedom to learn, and humanized education. I realized that I needed to be proactive about any dynamics of group work that may potentially produce power relations among students or reinforce existing ones. When students work in each other's ZPDs, I should be paying attention to the possible emergence of self-assigned leaders or exclusions. I decided to be especially proactive about situations in which peer tutoring may create a vertical relation between a more competent and a less competent student. A transmission style of education may occur when peers help each other.

Based on my observations of small-group work, I concluded that if ZPDs were applied to learning mathematics in the absence of an egalitarian community and dialogic classroom, it would run the risk of reproducing power relations inherent in traditional education. As Wells (1999) argued, changes created by ZPDs may transform or "reproduce existing practices and values" (p. 333). I realized that the ZPD concept, in the absence of a dialogic classroom, would lead to an oppressive and dehumanizing education, as described by Freire (2013).

Therefore, for each EUP, I facilitated a whole-class discussion in which the importance of dialogic peer interactions and egalitarian collaboration was emphasized. Students appeared to act upon these reflections as they worked in each other's ZPDs for any given activity. Thus, the ZPD concept became a catalyst for creating a more egalitarian community of learners. The reciprocal process was also evident: As the classroom became more egalitarian, students worked in each other's ZPDs more effectively. By building their mathematics knowledge on each other's ideas, thoughts, and problem-solving strategies, they seemed to make more progress than they could on their own. Regardless of mathematics ability, students asked questions, made comments, and responded to their peers' ideas. This finding is consistent with Cesar (1998), who concluded that if dialogic peer interaction is achieved in small-group work, a more competent peer is unnecessary to create a ZPD.

Empathy and Respect

The findings in my study revealed that empathy plays a significant role in fostering democratic peer interactions. Bin's journal entry for EUP 4 is worth revisiting:

> I saw that Jennifer's set of inequalities was wrong, but I did not say anything at first....You know, I did not want to make her feel bad....Instead, I asked questions to help our group to see one variable was missing in the system....I felt good about myself as I did not pose like a know-it-all jerk.

Bin knew instinctively that if he stated the correct answer immediately, it would make him look like a dominating person in the group, and make Jennifer feel bad. He consciously avoided this situation, choosing an empathic way instead.

Implicit in Bin's reflection is that he rejects the transmission style of education in favor of a facilitative style, which was modeled by the classroom teacher. This brings us to another conclusion about empathy. In a truly democratic community of collaborative learners, the teacher must be an empathetic facilitator, not an all-knowing authority figure. Needless to say, this requires a drastic shift from the assumptions of traditional education.

Forming Groups

In some EUPs, students formed their own groups. For example, in EUP 5, the students set their groups based on what they wanted to study in college: Those who wanted to study engineering formed one group; those who wanted to be nurses formed another. However, in EUP 3, I organized all the groups. I explained that to become a community—as opposed to individuals sitting next to each other in a room—we needed to socialize, work together, and get to know each other.

Years of teaching experience had led me conclude that students usually socialize with those from their own cultural background. At the planning stage of research, I considered this to be an obstacle to setting groups. I wanted students from diverse backgrounds to learn from each other.

Accordingly, at the beginning of each EUP, I explained that what unifies us as human beings is much stronger than what allegedly separates us, such as religion, race, or nationality. The students' journals and comments indicated their desire to consider our classroom a community. Although the existing literature, such as Kohn (1992), offers valuable ideas about collaborative learning in general, it provides few practical insights into how to organize group work. I ensured that each group included boys and girls, as well as students from different racial, religious, and ethnic backgrounds. For example, I did not put all the African American students in one group and all the Asian students in another.

These mixed groups seemed to work well. Overall, students worked in harmony; they respectfully listened to each other and responded to each other's concerns, suggestions, and ideas. Each group independently coordinated its work to complete any given EUP. The groups had some content-based questions, such as setting up or solving linear inequalities, for which I provided scaffolding to enable them to continue independently. I did not assign any member of a group a specific task to complete. On the contrary, the groups were free to decide for themselves how to complete the project.

Prior to each EUP, I presented the idea that students need skills such as developing initiatives and democratically working with others to frame an action plan for their projects. Such cycles of action-reflection reflect a fundamental goal of CME, which aims to help students become critical citizens. As others have noted (Aguilar & Zavaleta, 2012; Ellis & Malloy, 2007; Hannaford, 1998), mathematics education for critical citizenship should help students develop skills and attitudes such as working with others and exercising collective initiative.

In conclusion, findings from EUPs support the distinction between the collaborative learning of mathematics, which draws on communicative rationality, and traditional approaches, which draw on instrumental rationality. Based on classroom discussions, student journals, and my reflective journal, it can be concluded that there is a connection between the collaborative learning of mathematics and being an egalitarian community of learners. Collaboration is characteristic of an egalitarian community of learners. Collaborative learning grows along with a facilitative teacher attitude, dialogic peer interactions, and inquiry-based learning. One cannot be separated from the others. In the micro community that the classroom became, as students developed collective initiatives to complete each EUP, they experienced nothing less than democracy.

Critical Mathematics Education and Inquiry-Based Learning

Each EUP included inquiry processes that had two interrelated dimensions: content-based inquiry and critical mathematics literacy. In this study, the inquiry processes took distinctive forms based on concerns that CME brings to mathematics education. EUPs connected functional literacy (content knowledge) to critical mathematics literacy. Themes of EUPs that were derived from students' life seemed to enable the students to connect their personal experiences of sociopolitical issues to macro level sociopolitical problems.

Manconi, Aulls, and Shore (2008) suggested that inquiry can be framed within four domains: process, content, strategy, and context. Let us consider the findings with respect to these properties.

Inquiry as Process

My observations and students' reflections showed that the inquiry in each EUP was a process through which the students seemed to learn "generalizable process skills that are specific [to high school mathematics] but that carry broad transferability across many subject-matters" (Manconi et al., 2008, p. 249). In each EUP, the driving force of the inquiry process was the students.

In their journals, students reported that they developed transferable skills such as questioning techniques, listening to each other's ideas, making suggestions or comments, respecting classmates, trying to be open-minded, being inclusive, interacting without dominating each other, and being empathetic. As a consequence of cycles of reflection on their experiences, the quality of their engagement in inquiry increased. It is encouraging to compare this finding with Staples's (2007) finding that "students' participation and the opportunities to negotiate meanings seemed to influence their interpretations and understanding of practices in ways that facilitated their participation" (Staples, 2007, p. 37). Across EUPs, I consistently observed that as the students realized that the research was conducted not *on* them but *with* them, they took ownership of the collective inquiry process.

The students in each EUP articulated in their journals that working collaboratively enhanced their ability to integrate inquiry into word problems. In the process of inquiry, as the students worked in each other's ZPD, they exchanged ideas and revised their initial approaches to develop a collective investigation. The students indicated that this collaborative inquiry process helped them recognize some of mathematical misconceptions they had. Some students stated that these cycles of inquiry helped them realize that there are multiple ways of arriving at similar conclusions; open-ended mathematics problems may have multiple answers.

The students noticed that mathematics problems do not always have to lead to a set of absolute/correct answers; answers may take different shapes depending on one's ideological, political, and ethical perspective. For example, in EUP 2, the students solved the word problem both from an employee's perspective and an employer's perspective, achieving two different answers. EUP 4 was another example: although there was a single optimal numeric solution, each group arrived at it in a different way. I observed this point over time: The inquiry process helped students develop a sense of community in knowledge construction.

Inquiry as Content

EUPs required students to apply their numerical fluency, procedural knowledge, and conceptual understanding of mathematics to solve word problems. In this respect, the inquiry process helped students improve their content knowledge. I observed throughout EUPs that collaborative inquiry processes improved the students' conceptual understanding of content knowledge. For example, the students in EUP 4 applied systems of inequality to optimize limited resources. As the students applied procedural knowledge of multipart functions to Edward's job application, they seemed to understand the concept of functions better than they did in EUP 2.

Thus, an inquiry-based approach to learning mathematics, along with collaboration and dialogic pedagogy, seemed to help students develop conceptual understanding of mathematics content. This finding is consistent with that of Manconi et al. (2008): "Inquiry provides an opportunity to teach content at a deeper level and to apply knowledge" (p. 250). The finding also substantiated one of the objectives of CME, which is to "honestly and openly address...the instrumental life goals of the learners themselves, both in terms of needed skills and passing exams" (Ernest, 2002c, p. 6). The inquiry processes in my research served this goal. EUPs helped students improve their procedural knowledge, numerical fluency, and conceptual understanding of subject matter to be successful on traditional exams. In this sense, EUPs served both functional and critical literacy. At the reconnaissance stage, I anticipated that the students' lack of prerequisite content knowledge could be a potential barrier to effective collaborative inquiry. Therefore, prior to each EUP, I covered the mechanical aspects of related mathematical content to ensure each student in the class could comfortably solve mechanical problems. For example, we covered linear functions-equations prior to EUP 1; we processed graphical and algebraic analysis of functions prior to EUP 2; we reviewed systems of inequality and optimization prior to EUP 4, and exponential properties prior to EUP 5. Being proactive seemed to work. No student appeared to struggle with content knowledge in EUPs.

This outcome reflects the observation of Aulls and Shore (2008) that if students lack prerequisite content knowledge, they will not able to participate in collective inquiry. During EUPs, I noticed that the students were still improving their content knowledge and correcting their misconceptions. In this sense, the process of inquiry also helped the students to assess their content knowledge. I realized that providing students with prior knowledge is necessary, but it is not a linear process. The students continued improving their content knowledge while applying their knowledge in the context of EUPs.

As my journal indicates, content knowledge and application cannot be separated. Collaborative inquiry created an intellectually stimulating atmosphere where the students were collectively engaged in enhancing their conceptual understanding of the content matter. The students had to pose questions; explain suggestions, ideas, and arguments; and respond to each other's conclusions. For example, in EUP 2, finding the intercept point(s) of two functions was a mechanical task. However, as the x-coordinate of the intercept represented the number of hours per week that Edward worked at a part-time job, it became a meaningful concept.

Inquiry as Strategy

Although I designed and facilitated each EUP, I did not assign students to specific tasks, nor did I provide them with any prescribed method of inquiry.

In small groups, the students organized their own process, discussing how to handle, for instance, the division of labor and presentation of their final work. I observed that students gradually developed three strategies of inquiry. First, as all EUPs were designed as open-ended inquiries that challenged the traditional assumption that there is always a single and absolute answer, students gradually formed critical and strategic thinking skills to overcome the uncertainty inherent in open-ended problems. Second, the students collaboratively decided what to do next—a proactive approach that improved their organization and coordination skills. Third, although there was no antagonistic behavior in class, the students had disagreements from time to time; they appeared to develop strategies to resolve these conflicts. I realized that egalitarian collaboration and dialogic interaction played a vital role in resolving these disagreements and reaching joint decisions (Pine, 2009).

Thus students—instead of following a specified method—developed their own strategies for dealing with uncertainty, timetabling, and conflict, which allowed them to apply mathematics to various assigned problems. The findings in the current research are in line with Barfurth and Shore (2008), Manconi et al. (2008), and Wells (2009), who suggested that inquiry-based education enables students to develop and exercise problem-solving strategies.

Inquiry as Context

The results of my research suggest that context of inquiry is another distinctive element of inquiry-based learning within CME. The context of each EUP enabled the students to make connections between their life in the classroom (life-world) and a larger society (system); they, as a whole class, seemed to develop bottom-up responses to top-down neoliberal ideology and pedagogy.

This finding corroborates Manconi et al. (2008), who concluded that the context of inquiry should help students challenge dominant views in a given society and "construct a new understanding of the world" (p. 250). In light of this goal, I designed each EUP as a project-based inquiry with a theme derived from the students' life-world to enable them to relate their life in classroom to a larger sociopolitical system. I contextualized each EUP as a socioeconomic and sociopolitical issue. To promote critical literacy, I attempted to combine inquiry *into* mathematics and inquiry *with* mathematics. My inspiration was previous scholars who have discussed the importance of critical mathematical literacy—scholars such as Frankenstein (1983, 1990), Gutstein (2006), Skovsmose (2011), Skovsmose and Greer (2012), and Alrø, Christensen, and Valero (2010).

CME is unique in linking functional literacy with critical mathematics literacy. For example, in EUP 2, students used their functional literacy such as

130   CHAPTER 7

graphical and algebraic analysis of functions to develop a piece-wise function of Edward's job offer. However, as they extended the inquiry to determine which part of the function favors whom, they developed critical mathematical literacy.

As revealed in their reflections, themes of inquiry in EUPs helped students develop bottom-up responses to top-down neoliberal ideology. I concluded that a connection between the students' life-world and the system could be constructed through the contextualization of standardized curriculum content into word problems. Again referring to EUP 2, the students tried to help Edward with the job offer made by his employer. Using mathematical concepts such as linear and quadratic functions and graphs, systems of equations, and piece-wise functions, the students developed two multipart functions to model the job offer: one was in favor of Edward as employee; the other favored his boss. This context provided the framework for a whole-class discussion to understand our society as a class society. The students realized that we live in a class society; however, if we have critical mathematics literacy, we can make rational decisions and not be manipulated by others.

The findings suggest that EUPs were effective in connecting functional to critical literacy. However, it is worthwhile to briefly revisit EUP 3. I had hoped to invite students to an open forum to discuss the formatting power of mathematics (Skovsmose, 1994, 2011) and the historical development of mathematics. However, it became apparent that the context of inquiry and the objective of the lesson were too broad to materialize within two block periods. As a result, we only touched the surface of our objectives. I realized that the history of mathematics must be carefully integrated into each unit throughout the school year. The lessons learned in EUP 3 could provide the foundation for future studies to reframe the history of mathematics from the perspective of CME.

Obstacles to Inquiry

Findings from my classroom-based research highlight four main obstacles to inquiry-based learning in CME. First, there are few ready-to-use curricular materials that are compatible with CME. Second, the inquiry-based approach is time-consuming, which made it difficult to integrate into standardized and scripted curriculum. Third, the standardized curriculum limited the content flexibility of EUPs. Finally, I received no support from administration or my colleagues.

The second obstacle merits further discussion. Inquiry-based learning of mathematics is time-consuming. Aulls and Shore (2008) also noted that inquiry-based learning requires much more time than traditional approaches, such as direct teaching. As the neoliberal educational changes in the U.S. mandate covering certain curriculum content within a certain time, inflexible use of classroom time becomes a real barrier to inquiry in a mathematics class. I could have used a few more days for each EUP to expand the inquiry further. But due to time pressure, we only had two days for each EUP. This finding suggests that mathematics teachers who want to practice CME must allow for time constraints resulting from neoliberal educational impositions.

A lack of materials consistent with the concerns of CME was one of the major obstacles I encountered. Therefore, I had to develop EUPS on my own. As indicated in the literature review presented earlier, a project-based inquiry in CME should have a certain ethical and philosophical structure, as Freitas (2008) argued. Skovsmose (1994, 2011) also pointed out that critical mathematics education must distance itself from the official curriculum. This was a vital point in my research, as I aimed to make small openings for students to negotiate implications of imposed neoliberal educational implementations and counter neoliberal hegemony.

Recognizing this issue, Freitas (2008) suggested that revising word problems in mainstream textbooks could be the solution to the lack of curricular material for CME. Having developed my own EUPS, I concluded that, to some extent, revising word problems in mainstream textbooks could be helpful to integrating inquiry into mathematical content. However, I also realized that CME must be based on the context and themes derived from students' life-worlds, and from current social, political, and economic issues at the local, national, and global levels.

Many times during this study, I wished that there were a network of educators—Internet-based or otherwise—who are dedicated to CME both at the national and international level, so that I could take advantage of teachers' ideas and lessons that aim at functional and critical literacy in math. It would be great to look at an inquiry project of a math teacher from India, China, or elsewhere in the world. Similarly, it would be great for me to share my ideas with other math teachers and receive feedback to improve my inquiry projects.

As stated in my reflection, there is a need for ready-to-use inquiry-based projects. The current research provides a solid framework for developing project-based inquiry to counter the hegemony of neoliberal pedagogy. In this sense, EUPS can be considered an original contribution to CME. Furthermore, a network of critical mathematics teachers would enhance the quality of CME practices and provide a forum in which ethical, pedagogical, and philosophical aspects of project-based inquiry could be discussed.

However, the current findings also show that anyone who wishes to develop an inquiry-based project must take certain practical aspects into account. First among these is that mathematics teachers, at least in the U.S., must deal with common-core state standards (CCSS). Other contextual issues include student and local community characteristics, as well as current social, political, and economic matters (Schneider, 2015). The advantage of deriving themes of inquiry from students' social surroundings is that motivation and level of participation may be enhanced (Ball, Goffney, & Bass, 2005).

Certain outcomes of my research resonate with previous studies. For example, Staples (2007) claimed that "despite [the] compelling successes [of the inquiry-based approach], traditional models of instruction, particularly at the secondary level, still dominate the educational landscape" (p. 1). However, the studies mentioned by Staples tended not to probe the education system as the main source of dominance, but instead focused on the dynamics of isolated classrooms. Thus Staples (2007) concluded, "Teachers find it very challenging to organize and support student participation in these discourse-intensive practices that centralize students' ways of thinking....[The] staying-power of traditional models is mysterious" (p. 1).

Findings in my study contradict Staples's (2007) conclusion that the reason behind the dominance of traditional instruction is "mysterious." It is no secret that the transmission style of U.S. education, featuring memorization, recitation, and standardized assessment, is the result of neoliberal educational implementations beginning in the 1980s (Giroux, 2012; Hill, 2008). In her analysis, Staples identified the dominance of direct teaching as an obstacle to inquiry-based learning, without probing its root causes and political background: "The students had experienced traditional models of teaching and learning in middle school, and many reported that they did not enjoy mathematics nor math class...which also creates challenges for participation" (p. 34).

My research confirms Staples's (2007) point that students enter the classroom with already established values and attitudes toward certain ways of learning mathematics. However, this obstacle didn't come out of nowhere. Neoliberal implementations have shaped life in the classroom. At the reconnaissance stage of the current research, I realized that without acknowledging and challenging this fact, it would be very difficult to implement collaborative inquiry in a mathematics classroom under the current system, dominated as it is by neoliberal ideology.

To cope with this potential barrier to inquiry-based learning, I facilitated whole-class discussions that effectively turned our classroom into a public forum to discuss philosophical and educational implications of inquiry-based collaboration and dialogic learning. Through whole-class discussions, the students had opportunities to negotiate traditional education and critical education, based on their experiences in other classes and their experience during EUPs. As the students realized that this study took a critical approach to mathematics education and that they were equal partners in the process, they enthusiastically participated in EUPs. As an amendment to Staples's (2007) conclusions, the findings in the current research suggest that inquiry in mathematics education should enable students to critically reflect on their past learning experiences; inquiry-based learning should be collaborative and dialogic in order to cultivate critical literacy and promote critical citizenship.

Findings in the current research revealed that the inquiry process in CME is dialectically connected with egalitarian collaboration and dialogic interactions in the classroom. This makes sense, as inquiry is a *collective* activity that requires a community of learners. Inquiry can be framed as real-world problems to contextualize content learning of social, political, and economic issues. The context of inquiry enables the connection between functional and critical literacy in mathematics. Themes of inquiry derived from students' life-worlds help them understand and challenge oppressive implications of the existing social, political, and economic system on multiple fronts; whole-class discussions can be structured as self-assessments to keep improving the quality of the process.

Notes

1 This approach is also known as direct teaching or lecturing.
2 My experiences confirmed that students are usually afraid to share their answers, thoughts, and ideas when there is a single correct answer and learning is competitive.

CHAPTER 8

Critical Mathematics Education and Citizenship in the Neoliberal Era

This chapter provides an analysis and discussion of mathematics education that would enable students to develop certain values, attitudes, and skills to become critical citizens. I argue that an inquiry-based collaborative and dialogic teaching and learning of mathematics allow students to develop and practice democratic values and skills in classroom.

Critical mathematical literacy is central to discussions of mathematics education oriented toward citizenship and democracy. Dialogic pedagogy, collaborative learning, and inquiry-based instruction in secondary mathematics education have certain implications. This chapter explores ways in which collaborative and dialogic mathematics education can be facilitated to help students become critical citizens. It argues that neoliberal pedagogy promotes individually responsible consumer-based citizenship. Therefore, CME, as a counterhegemonic practice, should promote justice-based participatory citizenship.

Democracy in the Classroom

Inspired by Freire (1998), I developed EUPs to create a dialogic classroom environment for students to develop skills, values, and attitudes to become critical citizens. Pursuing this goal underscored the point that in order to cultivate democratic values, life in the classroom must be democratic. The combination of collaborative learning, inquiry-based approach, and dialogic pedagogy transformed our classroom into an egalitarian community of learners.

As anticipated, critical participatory action research (CPAR) resonated well with the objectives of CME. Exceeding my expectation, integrating inquiry into collaborative learning and dialogic pedagogy turned the learning process into a democratic forum and our classroom into a dialogic and facilitative environment where the students established a thick version of democracy. The students actively participated in decision making and inquiry-based collaborative learning process. Egalitarian and respectful peer interactions and horizontal teacher-student relationships became a self-regulatory process over time, turning our classroom into a community of mathematics learners.

© KONINKLIJKE BRILL NV, LEIDEN, 2019 | DOI:10.1163/9789004390232_008

EUPs created a communicative space for the students to push outside their comfort zone through inquiry and build knowledge on each other's ideas. While learning mathematics, the students seemed to develop and exercise skills and values, such as listening to peers, respecting different points of view, and empathizing with classmates. This finding is in line with those of Ellis and Malloy (2007) and Hannaford (1998), who argued that mathematics education oriented to promote critical citizenship should help students develop such skills and values.

Across EUPs, through cycles of action-reflection, the students seemed to make conscious attempts to transform the class to a dialogic classroom. Dialogic and collaborative learning were the defining elements in the process of the class becoming a community. This finding supports other research showing that education oriented toward a thick version of democracy and critical citizenship should promote dialogic learning and egalitarian collaboration (Gandin & Apple, 2002; Hyslop-Margison & Thayer, 2009). However, as my reflections and students' journals indicate, the collective learning in our class was not a matter of collective compliance. Each member in the group— actively and with equal power—participated in decision-making processes.

The EUPs together revealed three elements that enabled our classroom to become a democratic community. First, students developed a sense of belonging to the classroom community as well as a sense of responsibility and collectivism. This was an important step in countering neoliberal pedagogy, because neoliberal ideology advocates a market-based citizenship that promotes individual consumerism and omits communal solidarity (Hyslop-Margison & Thayer, 2009; Westheimer, 2015).

Second, collaborative learning enabled students to actively participate in knowledge construction and the meaning-making process. The students became agentic participants instead of passive consumers of knowledge. This way of participation can also be considered a means of countering neoliberal pedagogy and citizenship. The result is consistent with that of Freire (2000) and Gandin and Apple (2002), who argued that an educational process that envisions students as passive receivers of knowledge is oppressive and anti-democratic.

Third, in all EUPs, there was a dialectical connection between individual transformation and transformation of the whole classroom. Because the students did not see their peers as obstacles to their learning and success, they genuinely strived to learn from and with each other, resulting in both individual and communal growth. The individual growth in the current research can be considered a negation of the self-concerned individualism portrayed in neoliberal pedagogy (Abdi & Carr, 2013; Kohn, 1992; Rogers, 1995; Wells, 1999). In other words, the findings here show that if a classroom is democratic, then individual students' growth and the growth of the entire classroom resonate with one other.

In addition to these three implications of EUPs, the fact that small-group work was inclusive contributed to the process of the class becoming a community of mathematics learners. The students' proposal for non-dominating peer interactions in EUP 1 materialized inclusiveness as practical wisdom instead of an intellectual abstraction. On this point, Dewey (1916) indicated that attitudes and values could not be imposed: "The required beliefs cannot be hammered in; the needed attitudes cannot be plastered on" (p. 11). Complementing Dewey's point, Kohn (1992) suggested that students should be guided to internalize certain ethical values in order to make moral-based decisions. Whole-class discussions during EUPs seemed to provide the communicative space necessary for students to negotiate and consciously embrace certain values and virtues oriented toward being an egalitarian community of learners.

Hyslop-Margison and Thayer (2009) elaborated on this point: "Unless the desired behavioral traits are fully internalized into the value system of students, the character development objectives of thick democratic citizenship are unlikely to be realized" (p. 115). Based on student reflections, I realized that our whole-class discussions helped them revise and substantiate their views and values to make group work inclusive in a deeper, more meaningful way.

Our emphasis on values such as inclusiveness, equality, and non-dominating peer interactions brought about some practical results. Prior to this study, I had always had students with a grade of F or C who seemed to be alienated and outcast. However, after EUP 2, all students had either an A or a B. I concluded that egalitarian collaboration and inclusiveness notably increased equal distribution of mathematics education, which reflected on students' grades in traditional assessment as well. I think this result itself can be regarded as one indicator of our classroom being a democratic space.

It can be concluded that the class's being an egalitarian and inclusive community was a strong indicator that the students and I established necessary conditions for thick democracy; thus our classroom became a micro society. Our reflective actions (praxis) created small openings to democratize classroom life.

Democracy in a Form of Dialogue

In EUP 5, I co-constructed an exponential model to calculate interest rates. I facilitated this learning process by making numerical patterns visible for the students to make generalizations to construct the formula. The students built on each other's ideas and made most of the calculations with their graphing calculators. I structured the process based on mathematical axioms, properties, and theories. I did not continue to the next step without receiving the students' approval. In other words, when I justified my steps in this process, I drew not on my authority as the teacher, but on mathematical and logical verification.

I concluded that high school mathematics content and pedagogy can be dialogically structured and that such structure promotes critical citizenship. I was particularly inspired by Almeida's (2010) connection between mathematics and democracy, and Habermas's (1987) theory of the "ideal speech situation." According to Almeida (2010), just as government in a democracy must justify its actions by referring to the Constitution, mathematics teachers must justify their teaching through mathematical logic, axioms, and principles.

Habermas (1987) made a similar but broader case that non-dominating and democratic communication requires a verification process based on better argument, not greater power. As I attempted to apply this proposal throughout EUPs, I embraced my role as facilitator rather than as authority. Students' journal entries indicate that they considered the classroom a place where peer relations were egalitarian, the learning process liberating, and the classroom environment democratic.

Mathematical Literacy and Citizenship

One of the main objectives of EUPs was to promote critical mathematical literacy by giving each student an opportunity "to examine one's own and others' lives in relationship to sociopolitical and cultural-historical context" (Gutstein, 2006, p. 5). Being engaged in a dialogic pedagogy enabled students to interrogate social, economic, cultural, and political matters that affect their lives. EUPs seemed to help us create communicative spaces for the students to discuss, in EUP 1, standardized assessment, and develop a bottom-up response to top-down neoliberal impositions. In EUP 2, we challenged the dominant view of mathematics that problems in mathematics have always a single correct answer and that mathematics is politically neutral. In this context, we also discussed the importance of critical mathematical literacy for working-class citizens to make educated decisions to identify and protect their best interests. In EUP 3, we held an open forum to discuss collaboration and competition in terms of their implications for the common good and well-being of world citizens. In EUP 4, we approached community service and volunteer actions to counter neoliberal tenets in those domains. In EUP 5, we questioned the student loan debt crisis resulting from neoliberal educational policies and came up with foundational and short-term proposals to remedy the situation.

Findings in my research reveal a point that is consistent with the ideas of Nagda, Gurin, and Lopez (2003), who suggested that through action and reflection, "Students can develop a more abstract understanding of social life.... The permeable boundaries between the classroom and the larger world can allow students to continually reflect on their in-class learning in relation to the outside, and vice versa" (p. 169). Students connected their life in the class to that of a larger society by developing bottom-up responses to neoliberal top-down impositions. In this sense, EUPs served critical mathematical literacy and thus emancipation: each EUP, in addition to content-based empowerment, provided small openings to question the root causes of injustice and oppression.

EUPS provided evidence of the connection between critical mathematical literacy and critical thinking. For example, EUP 2 encouraged students to discuss labor relations and socioeconomic inequality in class-based societies. The students concluded that critical mathematics literacy is needed to know, communicate, and defend your best interests. Otherwise, you are vulnerable to distortion and manipulation, and possibly at the mercy of employers or others who are in a power position. To arrive at those conclusions, critical mathematical literacy involved critical thinking.

However, the term "critical thinking" is itself problematic, as it is also part of neoliberal discourse. Therefore, it is important to distinguish between critical thinking based on communicative rationality and critical thinking based on instrumental rationality. Neoliberal ideology drew on instrumental rationality to redefine critical thinking in market terms. In neoliberal pedagogy, critical thinking comprises a set of analytic thinking skills but no ethical or political orientation (Giroux, 1994). Critical thinking is considered only as individual action. Hyslop-Margison and Thayer (2009) explained: "These [neoliberal] critical thinking constructs promote technical [instrumental] rationality by encouraging students to address problems from a limited perspective that ignores wider workplace, labor market, and socioeconomic issues" (p. 27). In contrast, critical pedagogy takes a wider perspective, effectively transforming critical thinking into emancipatory praxis.

In EUP 2, students commented that Edward's story was interesting but unrealistic. Reflecting on their part-time job experiences, the students concluded that no employer would give such a choice to an employee, for several reasons: First, there is an asymmetrical power relation between workers and employers, and there is no democracy in the workplace that would lead to such an option for workers. Second, if we compare how much money an employee's labor brings to a company and the employee's salary, we realize that workers receive a small fraction of the money earned by the company. Therefore, from the students' perspective, the weekly salary in Edward's job offer was incongruent with reality of the current labor market. This shows that the students were engaged in critical thinking that draws on communicative rationality.

Based on the students' reflections, I realized that it is important to identify the neoliberal version of community volunteer service in order to design open-ended projects to counter such consumer-based citizenship. I therefore designed EUP 4 to put the concept of false generosity into perspective, thereby reclaiming community volunteer involvement for thick democracy and justice-oriented citizenship.

As I was designing this project, I exercised caution to avoid some neoliberal pitfalls. I realized that dealing with the notion of voluntary community service in an educational context was delicate. The goal was to provide students with an opportunity to negotiate democratic values such as solidarity and generosity in the context of helping homeless people, as well as to debate the root causes of poverty and homelessness in the U.S. The students' arguments and emotional responses against neoliberal tenets were far beyond my anticipation. In the context of Edward's story, the students commented on participation in voluntary organizations to help homeless people and contributed a reflective discussion to unpack social, economic, and political structures that produce homelessness and poverty in the first place.

In EUP 4, students used functional literacy—mathematics knowledge—to calculate the optimal number of rentals. They also debated Edward's possible motivation to participate in homeless shelter organization. Had we finalized the project at that point, EUP 4 would have served only functional literacy and promoted personally responsible participatory citizenship. In that case, the project would not have avoided the pitfall of false generosity.

However, we continued the project by discussing questions such as these: Why do social problems such as socioeconomic inequality, homelessness exist in our society? Is success/failure an individual or a social matter? Do we need values such as caring for others and solidarity or self-concerned individualism to establish and sustain a democratic and just society? In their responses, students radically challenged the premises of neoliberal ideology. The questions seemed to create a vivid intellectual and communicative ambiance that resulted in a dialogic classroom and promoted justice-based participatory citizenship (Orlowski, 2012; Westheimer, 2015).

EUP 4 was aimed at serving critical literacy and promoting democratic citizenship to counter consumer-based neoliberal citizenship. High school students in the U.S. are often involved in community volunteer services. However, they tend to consider such service as a personal investment to help build a strong resume for their college application. A consumer-based neoliberal pedagogy discourages students from perceiving community service in terms of citizen responsibility and solidarity (Brown, 2015; Lund & Carr, 2008).

Similarly, Westheimer (2015) pointed out that participation alone does not lead to a justice-oriented involvement. Westheimer emphasized the significance of critical literacy for justice-oriented involvement in volunteer services. Freire (2000) also indicated that a community volunteer service or any civil organization aimed at helping people in need should avoid the pitfall of false generosity. Westheimer's (2015) example of students' involvement in a food drive clarifies this concept. If students participate in a food drive but are not provided with the necessary tools to investigate the root causes of poverty and hunger, it actually helps reproduce the existing system while giving students the illusion that something beneficial is being done.

Critical Pedagogy and Critical Citizenship

The intimate relationship between mathematics education and democracy necessitates an analysis of related concepts such as "critique" and "citizenship," together with classroom practices that involve dialogue as pedagogy, inquiry, and collaboration. These concepts need to be interconnected in order to find answers to the question, How should mathematics classrooms be conceived such that students become active and critical citizens? This need comes from the fact that neoliberal ideology, drawing on positivism, proposes its own definition of democracy: It aims to reshape the education system to produce "instrumental citizenship" in support of a market-driven society. Critical citizenship must therefore counter the cultural hegemony of neoliberalism. Reviewing John Dewey's ideas about the connection between democracy and education is a suitable point of departure to distinguish a critical theory of education from a progressive theory, and then to elucidate the standpoint of critical theory on neoliberalism.

Dewey's (1916) emphasis on the connection between democracy and education may help us understand the notion of critique in the historical tradition of critical theory. From Dewey's perspective, democracy can be best built and maintained through the scientific method, as it rejects all dogmatic approaches as well as external authorities. In this view, the scientific method is the motor of progress both for individual and social life. Education should provide students with opportunities of inquiry-based learning to internalize scientific formation, which will eventually lead to unfettered thinking and progress in society; this process will eventually help students develop citizenship skills and integrate them into the society. This Dewey-inspired perspective uncritically embraces scientific rationality based on the ideals of the Enlightenment.

However, sociopolitical experiences such as rise of fascism in continental Europe and the Stalinist movement in Soviet Russia in 1930s obligated members of the Frankfurt school to critically review the ideals of the Enlightenment, including scientific rationality. A member of the Frankfurt school, Marcuse (2013), argued that the scientific method (positivism) may not always be progressive and lead to free thinking. In fact, it could become an oppressive force that reduces human beings to a single dimension; therefore it should be subject to critique. A member of the second generation of the Frankfurt school, Habermas (1972), addressed Marcuse's criticism. He argued that modernism is an unfinished project that includes not only scientific-technical knowledge interests, but also "practical" and "emancipatory" knowledge interests.

Overall, critical theory advanced an interdisciplinary project that aims not only to explain existing conditions but also to change them. As Skovsmose and Alrø (2004) argue, Dewey's view of education in relation to democracy is non-critical. Therefore, CME should transcend Dewey and show "how critique and learning mathematics can be connected and how the competence of mathemacy can be supported" (Skovsmose & Alrø, 2004, p. 20) in order to help students develop and experience critical citizenship.

A citizen in a general sense is someone who is a legally recognized member of a state: This recognition entails certain rights and responsibilities. However, one must possess certain competencies and be willing to utilize them to move from being a passive-obedient citizen to an active, critical citizen. Citizenship in this sense has cultural implications extending far beyond its legal definition. As Stevenson (2011) frames it, "By citizenship, I mean our connection to particular social and cultural locations, the possibility of a participatory involvement in shaping our society, and our understanding of our rights and responsibilities" (p. 5). In the U.S., this definition notably includes children of immigrants, whose parents have no legal ties to the government, and who are often regarded as "undocumented or illegal immigrants."

Education in a given society usually aims to reproduce the status quo; schooling practices are designed accordingly. The curriculum in most countries contains "citizenship" or "civic" courses in which a good citizen is considered

> Someone who possesses certain propositional knowledge, often including rather banal facts about national history and electoral/legislative process, and applies this information in a certain prescribed fashion. Such as citizen might be involved in community service, obeys legal dictates, and feels compelled to cast a ballot when civic duty affords the opportunity to do so. (Hyslop-Margison & Thayer, 2009, p. 1)

This type of citizen can only reproduce the status quo. Critical citizenship goes far beyond. For example, a critical citizen may have skills of adaptation to society as well as competencies to participate in collective initiatives that challenge oppressive, exploitive, and undemocratic situations.

Doganay (2012) points out that citizenship education does not have to be delivered in a separate civics course. Instead, it should be integrated into all courses and at all grade levels. Any course can help students develop skills, values, attitudes, and knowledge to become active and democratic citizens. Beutel (2012) argues that lessons and projects designed as group work can promote some of these qualities: "Even the project itself is a democratic method by means of its basic elements: choosing and planning a topic jointly, realizing it with all group members and recording as well as evaluating it together" (p. 12). Simmt (2001) contrasts instructional practices that contradict citizenship and those that promote active citizenship. Mathematics instruction that presents mathematics as "a set of facts, skills and procedures" oriented toward a single right answer does not promote citizenship qualities at all. However, instruction that embraces a problem-posing approach and applies investigation and dialogue would potentially promote qualities of good citizenship.

After reviewed the neoliberal threat, Hyslop-Margison and Thayer (2009) outline qualities of critical citizenship in the classroom from critical theory and pedagogy, including the views of Habermas and Freire. A critical citizen

- can form individual and collaborative initiatives and "participate in their community by volunteering or engaging in various forms of political activism and they educate others around them on the issues";
- possesses communicative competency to be engaged in dialogue with others to reach consensus "rather than dismissal simply on the grounds of initial disagreement";
- has intellectual and civic courage to listen to and review different ideas and revise his/her view not "on the basis of external pressure or political isolation but change their views only in the face of justifiable arguments and/or evidence against them";
- should be searching, with a "healthy skepticism" and critical standpoint, "new knowledge, reflecting on that knowledge and translating it into immediate social action";
- is critically literate not to be manipulated by hegemonic forces and able to distinguish things that can be changed such as social, economic, and political structures in a society from things that are given. (pp. 117–118)

While Hyslop-Margison and Thayer (2009) proposal is inspirational—it coherently frames what a critical citizen should be able to do—it omits a very important question: How can these objectives be realized given neoliberal restrictions? The goal of this book has been to address that question.

Neoliberal Hegemony and Mathematics Education for Critical Citizenship

The findings from my research indicated that EUPs promoted democratic mathematics education and critical citizenship. EUPs
- Allowed students to be active co-creators of knowledge;
- Cultivated a culture of collaborative learning and collectivity;
- Promoted inquiry-driven education as opposed to skill-drill types of exercise;
- Transformed the class from crowd of individual students to a community of learners;
- Promoted critical as opposed to functional literacy;
- Promoted justice-based participatory citizenship as opposed to consumer-based market-driven individual citizenship.

However, the findings from my research also identified certain obstacles to CME. Two stand out. First, neoliberal educational implementations affect the entire school atmosphere. The second is a limitation that concerns the community of critical educators.

During EUP 3, the principal visited my classroom to do a formal evaluation, and expressed concern that my lesson was not aligned with the standardized curriculum. I responded to the principal's question by explaining the importance of the students developing insight into the historic dimension of mathematics. The principal later told me in person that I should have covered nothing but the standardized curriculum in my lessons. This particular situation did not result in any serious consequences—only a mild warning. However, it could have negatively impacted my professional evaluation and even threatened my job security. This event made me anxious for the rest of the project: I kept asking myself, What if the principal shows up during a discussion of poverty, homelessness, or the student loan debt crisis? I would have to provide a standardized learning target for the lesson! I had no short answer for the question, as the common-core state standards do have not a single reference to democracy or citizenship. In that case, I would have violated the administrator's expectation twice, and a more serious consequence would be likely.

I took the risk, but limited whole-class discussion to one day for each project. Practicing CME in a classroom colonized by top-down neoliberal pedagogy necessitates a course of action to create small openings where bottom-up responses could be initiated. I realized that creating small openings was riskier than I had anticipated at the beginning. Neoliberal education system has its own version of pedagogy, democracy, and citizenship that shrinks space for other alternatives. Under these circumstances, practicing CME became more difficult and risky than ever. On the one hand, my commitment to CME provided me with moral energy and hope to say, "I should continue subversive teaching no matter what." On the other hand, to be able to teach, I needed to keep my job. This was a more delicate issue than I thought.

While I was conducting the project, I attempted several times to discuss my experiences during staff meetings and professional development days. Each time, the school administrator rejected my request by claiming that the topics of discussion were inconsistent with their agenda and, therefore, would be disruptive. I was disappointed that I was not allowed to share my research with my colleagues! Our administrators were engaged in public relations of "classroom-based research" sponsored by big corporations and think tanks. But they were not interested in a research study conducted in their own school. I cannot blame individuals in the administration: they were expected to follow a scripted agenda mandated by the system, which exploits the concept of research as an apparatus to justify market-driven changes in education.

I knew that educational ideas critical of neoliberal educational policies and implementations are not welcomed. However, my original thought was this: given that the standardization movement in education has been presented as research-based "best practice," the school district or at least the principal would be interested in the findings of my research as it was conducted in our school. On the contrary, I was not allowed to share my research. This experience confirmed Apple's (2000) conclusion that the neoliberal managerial approach is "not based on trust, but on a deep suspicion of the motives and competencies" (p. 70). I received no feedback or recognition for inquiry-based collaborative and dialogic teaching of mathematics in my class.

As indicated earlier, my research was hampered by a lack of curricular materials and professional development opportunities. Designing EUPs on my own was a time-consuming activity that increased my workload. This issue cannot be blamed on neoliberal colonization. However, the CME community1 as a whole has to face the problem, as it limits the practicality of CME in classrooms. Studies in CME and critical pedagogy offer very limited practical tools for their classroom application. I wish that there were similar studies done by other math teachers or researchers that I could use to develop my own projects. What can be done? I think there could be a network, perhaps international, of critical math teachers to exchange ideas and classroom experiences. This would challenge critical educational theorists who undertake theoretical studies in critical pedagogy and CME to focus on life in classrooms. It would address this issue by restoring the dialectical connection between theory and practice as well as opening up critical teachers' vision to find ways of practicing CME and critical pedagogy under the restrictions of standardization.

CME AND CITIZENSHIP IN THE NEOLIBERAL ERA

In the process of developing my own projects, I noticed that this issue reflects more than a lack of curricular material. Having reviewed literature in critical pedagogy and CME, I realized that researchers from these communities have no clear information about teaching and learning practices in high school classrooms in the U.S. For example, the CME literature includes very few studies about CME for democracy and citizenship—most of the research is theoretical. As these theoretical studies are produced without feedback from life in classrooms, they become, of necessity, theories about theories.

Due to a lack of dialectical connection between theory and practice, the language of critical pedagogy and CME tends to become abstract and opaque. I noticed that critical educational researchers argue that neoliberal hegemony has eradicated possibilities of an education that would promote a thick version of democracy and critical citizenship. However, at the same time, they enthusiastically champion CME and critical pedagogy as more important than ever. But if neoliberal ideology has truly eliminated the possibility of critical pedagogy and CME, the recommendation to practice it would seem to be nothing other than empty rhetoric.

A leading figure in critical educational research, Michael Apple (2013), recognized this problem, arguing that critical scholars "have not been sufficiently connected to the actual realities of schools and classrooms" (pp. 50–51). A researcher in CME, Ernest (2010) used stronger language when he questioned academic work. He argued that scholars who write articles on CME without having a connection to classroom experiences actually are "complicit with the [neoliberal] system" that they are so critical of (p. 82). Blacker (2013) agreed, claiming that scholarly articles written by critical educators and theorists may improve the writers' careers, but they do not help resist neoliberal colonization.

Thus there is widespread agreement that the gap between theory and practice presents a serious limitation for CME. I hope that my research goes some way to bridging it. To the extent that it does, this study opens new domains for research aimed at expanding CME from the classroom to school and community to counter neoliberal hegemony and generate hope in the possibility of a more just and rational world.

Note

1 By CME community, I mean educators and writers who are committed to critical pedagogy and produce academic work on critical (mathematics) education.

CHAPTER 9

Conclusions

My main objective in this study has been to investigate the scope and limitations of critical mathematics education (CME) in the neoliberal era. CME is a relatively new but growing research field. CME to date comprises mostly theoretical studies—there is a serious shortage of classroom-based research. Indeed, my research is the first classroom-based critical participatory action research in CME in a high school context. With this book, I aim to bridge the gap between theory and practice in CME.

I conducted the study in a public high school mathematics classroom in the U.S. to answer the central question:

– What are the potentials and limitations of CME in terms of classroom teaching in the neoliberal era?

The findings led to the conclusion that despite an educational environment resulting from the market-based standardization movement, CME can be implemented through the interconnected dynamics of collaborative learning, dialogic pedagogy, and inquiry-based practice. When these elements were oriented toward promoting critical citizenship and a "thick" version of democracy, students began to take on democratic values, critical mathematical literacy, and critical citizenship.

The main conclusion of my research concerns three domains. First, the cycles of plan-act-observe-reflect gradually turned the classroom into an egalitarian community of mathematics learners. A facilitative pedagogic ambiance was created in which the students experienced mathematics learning in the form of a dialogue. Second, lessons presented as end-of-unit projects (EUPs) created a communicative space for students to develop and exercise critical mathematics literacy, democratized the learning process, and initiated bottom-up responses to counter the hegemony of neoliberal ideology. Third, certain practical limitations of CME must be acknowledged, given the overwhelming neoliberal colonization of education in the U.S.

Dialogue, Collaboration, and Inquiry

Findings from my research allow us to reinterpret the existing literature to distinguish dialogic pedagogy in CME that draws on communicative rationality

© KONINKLIJKE BRILL NV, LEIDEN, 2019 | DOI:10.1163/9789004390232_009

CONCLUSIONS

from other dialogues that draw on technical (instrumental) rationality. Dialogic pedagogy in CME is egalitarian, which is crucial for an inclusive process of teaching and learning. In this connection, recall that students' first proposal was to insist on non-dominating peer relationships in group work. Dialogic pedagogy in CME is structured through inquiry-driven learning materials to create communicative space in the classroom. To be motivational, these materials must reflect students' lived experiences; because they focus on significant issues, they act as catalysts to open spaces for unpopular ideas to be considered.

There is currently a gap in the CME literature concerning classroom practice. The present findings show that collaborative learning provided students with a more equal access to mathematics knowledge and skills, and also helped them develop a sense of belonging necessary to promote a thick version of democracy. As the collaborative process contradicts neoliberal pedagogy, it requires a radical shift from market-driven, individualist, competitive learning to egalitarian and facilitative learning processes. My students and I materialized this shift through whole-class discussions. After reflecting on previous experiences with group work, students unanimously decided in favor of egalitarian peer relationships. Thus they rejected the political, pedagogical, and ethical consequences of individualistic competitive learning.

The inquiry process in CME must be in harmony with dialogic pedagogy and collaborative learning. Across EUPs, students became more agentic participants. They collectively negotiated and constructed knowledge; nothing was imposed on them. This way of learning mathematics subverted the transmission style of education and served as our response to top-down neoliberal pedagogy. Freire (2000) elucidated such an approach when he noted, "Projecting an absolute ignorance onto others, a characteristic of the ideology of oppression, negates education and knowledge as processes of inquiry" (p. 58). Confirming Skovsmose's (1994, 2011) argument that inquiry in CME replaces the exercise paradigm and rote memorization, the current study showed that CME can be realized through open-ended word problems, where the goal is for students to develop transferable skills, improve content knowledge, and contextualize mathematics into sociopolitical and socioeconomic issues. While we were engaged in a practice of CME, the students still had to pass standardized tests. The inquiry process, therefore, cannot neglect the content knowledge that students will need for the test.

The Mathematics Classroom as a Micro Society

Although some theoretical studies emphasize the importance of making the classroom a community (Kennedy, 2009; Murphy & Fleming, 2010), none of them concerns mathematics. Put differently, the CME literature is silent on classroom-based approaches. The most important original contribution of my study to CME, therefore, is that it is firmly rooted in an actual U.S. high school mathematics classroom. The present study bridges the gap between theory and practice, because a mathematics classroom was transformed into a community.

A central tenet of CME is that all classroom practices must be oriented toward creating an egalitarian community of learners. The basic elements of CME in the classroom—dialogic pedagogy, collaborative learning, and inquiry-based lessons—are unsustainable if they are not dialectically structured to establish and maintain an egalitarian community. The following three interconnected sub-conclusions substantiate the central conclusion.

First, the present findings show that mathematical concepts can be taught through dialogic pedagogy—authoritarian teaching is not the only way. The CME literature distinguishes between dialogical and nondialogical teaching of mathematics. I posed the following question: Can CME completely avoid nondialogical (authoritarian) teaching? Disproving Mortimer and Scott's (2003) claim that the authoritarian approach is inevitable when mathematics and science teachers introduce a new topic, the findings here indicate that dialogic teaching is effective for teaching mechanical aspects of mathematics. However, introducing a topic through dialogue is not attainable in a traditional classroom driven by vertical student-teacher relations: It requires an egalitarian community.

Second, this study confirmed that students' learning improved to the extent that they were able to learn from and with each other to materialize their full potential (Vygotsky, 1978); there was no need for more competent students in small-group work. This process of egalitarian peer collaboration also helped me as the classroom teacher to become a facilitator (Wells, 1999). These findings have an important implication for the notion of ZPD: In order to apply ZPD as part of CME practice, the classroom must be an egalitarian community of learners. In the absence of a facilitative classroom environment and egalitarian peer interactions, the ZPD process could instead produce power relations among peers, thus reproducing transmission-style education.

Third, my research revealed that the "ideal speech situation" outlined by Habermas (1990, 2005) can be attainable only if the classroom is an egalitarian community of learners. After four cycles of planning, acting, observing, and reflecting (EUPS 1 to 4) our classroom had visibly become an egalitarian community. Creating conditions for the ideal speech situation was a time-consuming process and required a radical change in power dynamics. However, we were rewarded with qualitative changes in peer interactions and student-teacher relationships.

CONCLUSIONS

The classroom transformed into a much more inclusive and a more open communicative space, especially in EUPs 4 and 5, where no coercive forces were in place to motivate students. Each student had equal power to question, speak, and act in small-group work and whole-class discussions. There were no secret agendas or distorted communication; the driving forces of interaction among the students were mutual trust, honesty, and respect. Nothing was imposed on students: Meaning making was driven by argument, not by rank. Findings also demonstrated that empathy is a significant element in creating space for ideal speech situations. Empathy in the classroom cannot be generated in the absence of love and hope, as articulated by Freire (2000).

Citizenship and Mathematics Education

As there is no previous classroom-based research on CME linking mathematics education to democracy and critical citizenship, this study provides the first response to the question. The answer can be framed in four parts.

First, inquiry-driven collaborative learning and dialogic pedagogy democratized life in the classroom. In EUPs, the students experienced mathematics learning as a democratization of classroom life. We experienced a "thick" as opposed to a "thin" (neoliberal) version of democracy (Orlowski, 2012; Westheimer, 2015). In agreement with Freire (1998), I found that teaching critical citizenship necessitates democratizing life in the classroom. As the classroom became a democratic space, we had a solid ground to relate mathematics to larger social, economic, and political issues.

Second, the study revealed the significance of making small openings in the classroom colonized by neoliberal (and neoconservative) educational implementations. The openings created by EUPs allowed me to incorporate critical mathematical literacy and critical thinking into the standardized curriculum. The students discussed some premises of neoliberal ideology and questioned irrational and unjust implications of market-driven educational policies. Through whole-class discussions, the students developed a collective, bottom-up response to neoliberal hegemony. In their view, education is a human right and a social investment, not an individual commodity and personal investment. As they embraced inquiry-based collaborative and dialogic learning, they rejected the competitive, authoritarian, and rote aspects of neoliberal pedagogy.

Third, the process of developing bottom-up responses entailed critical thinking as part of critical literacy. Engaging in the structural analysis of society and imagining a better one, the students objected to corporations' involvement in

150   CHAPTER 9

education and made proposals to make society at both micro and macro levels more just, equal, and sustainable. In this sense, the students were engaged in critical thinking that draws on communicative rationality and that recognizes the ethical and political dimensions of critical thinking. This version of critical thinking differs radically from the neoliberal version that draws on technical rationality to solve business problems. EUPS prompted critical mathematical literacy, through which the students developed the ability to question authorities and keep them accountable. Therefore, this study promoted a thick version of democracy and a participatory, social justice-oriented citizenship.

Fourth, the students clearly opposed neoliberal policies and implementations. When communicative space was made in the classroom, students raised their voices against the neoliberal world view. Students' journals indicated that they did not consent to neoliberal ideology as a dominant discourse. As Habermas (1975) argues, the system colonizes the life-world and prevents free public debate, which makes the legitimacy of the system questionable. The legitimacy issue applies to educational policies as well. The results here show that a start can be made by creating small openings in the classroom, where students can develop bottom-up responses to counter neoliberal colonization.

Curricular Materials in CME

There is a gap between theory and practice of CME in terms of developing word problems and projects to counter neoliberal pedagogy. The findings from the current study clearly begin to bridge the gap. In relation to the elements of CME, five aspects of potential projects and word problems can be identified.

First, a distinctive element of dialogic pedagogy in CME is alternative learning materials; notably, open-ended problems and projects. Projects must be in harmony with the principles of dialogic learning. The present findings corroborate Skovsmose's (1994) suggestion that problems in CME should create a domain of investigation. Problems must be a forum where students can relate their learning to a larger society in order to negotiate social, political, cultural, and economic issues that affect their lives. I conclude that CME projects must not be limited to the exercise paradigm or solving a modeling problem.

Second, word problems and projects in CME must be multilayered, so that students working in small groups can negotiate implications of the problem and build on each other's contributions.

Third, problems must be inquiry-driven for the students to learn mathematical skills and knowledge that are transferable to different domains of study. Word problems must help the students improve their conceptual

CONCLUSIONS

151

understanding, procedural knowledge, and numerical fluency in order to pass standardized tests and be successful in the conventional sense.

Fourth—and the most important element of problems oriented toward CME—they must be built on clear ethical and political grounds to be able to counter neoliberal hegemony. It is worthwhile here to revisit EUP 4. We contextualized the community volunteer service issue to counter neoliberal hegemony in education. I noticed the significance of the political and ethical ground on which I designed the project to distinguish the notion of helping others—"false generosity" (Freire, 2000), in a thin version of democracy—from solidarity in a thick version. The same distinction can be applied to the notion of critical thinking: To promote critical mathematics literacy, word problems and projects should help students distinguish between critical thinking based on communicative rationality and one based on technical rationality.

Fifth, the findings serve as a reminder that critical mathematics teachers need to be aware of risks to their job security and be proactive about them. The learning targets in the U.S. standardized curriculum are part of the management and control process in public schools. A mathematics teacher, therefore, must find ways of linking word problems and projects to these standards. Otherwise, they could face disciplinary consequences. In my case, each EUP, with one exception, was connected to a specific learning target outlined by the school district. However, I could not link EUP 3 to any learning target, as the standards do not mention the history of mathematics. The principal's classroom visit at that time put a question mark on my evaluation. This caused only a minor problem for me, but it could have turned into a much more serious issue.

The conclusion is that for the sustainable practice of CME, word problems and projects must be linked to learning targets in the standardized curriculum. This is a new contribution of my study to the existing CME literature. However, I do not claim that my conclusion is the final answer. There is a need for more classroom-based research from different parts of the country—and from other countries—to provide political and pedagogical insight into integrating word problems and projects into the standardized curriculum without punitive consequences.

Micromanagement and Control

This study has also shown that a mathematics teacher who wants to practice CME should allow for some possible consequences. CME is not welcomed in schools colonized by neoliberal pedagogy. Although one may succeed in creating an egalitarian community of learners in the classroom, life in other classes is mostly organized by market-driven educational discourses. This situation could demoralize students and teachers alike. Therefore, a practice of CME must openly negotiate these kinds of situations with students through whole-class discussions.

My research study shows that it is possible to practice a humanizing education that sides with students as human beings and citizens against the imperatives of the neoliberal system; however, such a practice is accompanied by certain political challenges. It is sustainable only if the classroom is treated as a democratic community. Because market-driven objectives currently colonize classroom life, it is imperative to create small openings in which students can develop critical mathematical literacy, reclaim their voices, and thereby subvert neoliberal hegemony.

Participatory Action Research and Critical Mathematics Education

This study drew on critical participatory action research (CPAR), which enabled us to open "communicative spaces" (Kemmis, 2008, p. 126) in the classroom to initiate bottom-up responses to top-down educational policies. Cycles of plan-act-observe-reflect were in harmony with the natural flow of classroom teaching; students connected learning to sociopolitical and socioeconomic problems as equal participants in the research process.

The current study supports Skovsmose and Borba (2000), who suggested that participatory action research that focuses on changes in the classroom resonates with the concerns of CME. This study showed that CPAR resonates well with an inquiry-based collaborative and dialogic pedagogy. CPAR enabled us to democratize classroom life; it proved to be an effective instructional model to practice CME. However, further classroom-based studies are necessary to be able to develop a framework for sustainable instructional practices through CPAR.

Limitations and Suggestions

My study provides a solid framework for CME in relation to market-driven educational changes. However, I do not claim that this is a complete frame. I conducted research in one class in one high school. Although neoliberal educational policies have been widely implemented across the U.S., educational changes may have impacted states—even school districts within states—to different degrees. Therefore, further research in other settings

CONCLUSIONS

should be undertaken to develop a more comprehensive picture of the scope and limitations of CME.

Final Word

Neoliberal ideology, pedagogy, and culture still exercise hegemonic power at the global level. Although there have been some successes at the local level, political movements against neoliberalism in different parts of the world have not yet created a viable alternative; they are still in a defensive stage.

I am aware that the transformative changes we made in our classroom may not mean so much at the macro level. Nevertheless, this study created small openings in a high school classroom and initiated an egalitarian community of mathematics learners. By doing so, it showed that a classroom *could* be transformed into a community and thus neoliberal pedagogy could be countered. With the creation of many more small openings, these promising results could be expanded to show that a dialogic teaching of mathematics and a more democratic education is possible, even within educational conditions that are hostile to the larger emancipatory vision of critical mathematics education and critical pedagogy.

References

Abdi, A. A., & Carr, P. R. (Eds.). (2013). *Educating for democratic consciousness*. New York, NY: Peter Lang.

Abrams, S. E. (2016). *Education and the commercial mindset*. Cambridge, MA: Harvard University Press.

Aguilar, M. S., & Zavaleta, J. G. M. (2012). On the links between mathematics education and democracy: A literature review. *Pythagoras, 33*(2), 1–15.

Alexander, R. J. (2005). *Culture, dialogue and learning: Notes on an emerging pedagogy*. Paper presented at the International Association for Cognitive Education and Psychology (IACEP), University of Durham, Durham.

Alexander, R. J. (2006). *Towards dialogic teaching: Rethinking classroom talk*. Cambridge: Dialogos.

Alexander, R. J. (2015, May). *Dialogic teaching essentials*. Retrieved from https://www.nie.edu.sg/files/oer/FINAL Dialogic Teaching Essentials.pdf

Almeida, D. F. (2010, May 15). Are there viable connections between mathematics, mathematical proof and democracy? *Philosophy of Mathematics Education Journal*. Retrieved July 12, 2014, from http://socialsciences.exeter.ac.uk/education/research/centres/stem/publications/pmej/pwome25/D. F. Almeida Are There Viable Connections.docx

Alrø, H., Christensen, O. R., & Valero, P. (2010). *Critical mathematics education: Past, present and future: Festschrift for Ole Skovsmose*. Rotterdam, The Netherlands: Sense Publishers.

Apple, M. (2000). Between neoliberalism and neoconservatism: Education and conservatism in a global context. In N. C. Burbules & C. A. Torres (Eds.), *Globalization and education* (pp. 58–77). New York, NY: Routledge.

Apple, M. (2005). Doing things the "right" way: Legitimating educational inequalities in conservative times. *Educational Review, 57*(3), 271–293.

Apple, M. (2013). Creating democratic education in neoliberal and neoconservative times. *Praxis Educativa, 17*(2), 48–55.

Apple, M. (2014). *Official knowledge: Democratic education in a conservative age*. New York, NY: Routledge.

Aulls, M. W., & Shore, B. M. (2008). *Inquiry in education: The conceptual foundations for research as a curricular imperative* (Vol. 1). New York, NY: Lawrence Erlbaum Associates.

Bakan, J. (2011). *Childhood under siege: How big business targets your children*. New York, NY: Free Press.

Ball, D. L., Goffney, I. M., & Bass, H. (2005). The role of mathematics instruction in building a socially just and diverse democracy. *The Mathematics Educator, 15*(1), 2–6.

Ball, S. (1993). Education markets, choice and social class: The market as a class strategy in the UK and the USA. *British Journal of Sociology of Education, 14*(1), 3–19.

Barfurth, M. A., & Shore, B. M. (2008). White water during inquiry learning: Understanding the place of disagreements in the process of collaboration. In B. M. Shore, M. W. Aulls, & M. A. Delcourt (Eds.), *Inquiry in education* (Vol. 2, pp. 149–164). New York, NY: Lawrence Erlbaum Associates.

Beutel, W. (2012). Developing civic education in schools. In D. Lange & M. Print (Eds.), *Schools, curriculum and civic education for building democratic citizens* (pp. 7–17). Rotterdam, The Netherlands: Sense Publishers.

Blacker, D. (2013). *The falling rate of learning and the neoliberal endgame.* Washington, DC: Zero Books.

Boyles, D. (1998). *American education and corporations: The free market goes to school.* New York, NY: Garland.

Brantlinger, A. (2014). Critical mathematics discourse in a high school classroom: Examining patterns of student engagement and resistance. *Educational Studies in Mathematics, 85*(2), 201–220.

Brown, W. (2015). *Undoing the demos: Neoliberalism's stealth revolution.* London: MIT Press.

Carr, P. R., & Porfilio, B. (Eds.). (2015). *The phenomenon of Obama and the agenda for education: Can hope audaciously trump neoliberalism?* Charlotte, NC: IAP-Information Age Publishing.

Carr, W. (1995). *For education: Towards critical educational inquiry.* Buckingham: Open University Press.

Carr, W., & Kemmis, S. (1986). *Becoming critical: Education, knowledge, and action research.* London: Falmer Press.

Cesar, M. (1998). *Social interactions and mathematics learning.* Paper presented at the International Conference: Mathematics Education and Society, University of Nottingham, Nottingham. Retrieved from http://www.nottingham.ac.uk/csme/meas/measproc.html

Chomsky, N. (2003). *Chomsky on democracy and education.* New York, NY: Routledge.

Chomsky, N., & Herman, E. S. (2008). *Manufacturing consent: The political economy of the mass media.* New York, NY: Pantheon Books.

Chubb, J. E., & Moe, T. M. (1990). *Politics, markets, and the organization of schools.* Washington, DC: The Brookings Institution.

Coco, L. E. (2013). Mortgaging human potential: Student indebtedness and the practices of the neoliberal state. *Southwestern Law Review, 42*, 565–603.

Cornelissen, L. (2014). *"The finest of neoliberalism's tricks." Why neoliberalism and democracy are wholly incompatible.* Paper presented at the Neoliberalism and Everyday Life: 9th Annual International Interdisciplinary Conference, Centre for Applied Philosophy, Politics and Ethics, University of Brighton, Brighton.

REFERENCES

D'Ambrosio, U. (1999). Literacy, matheracy, and technocracy: A trivium for today. *Mathematical Thinking and Learning, 1*(2), 131–153.

D'Ambrosio, U. (2010). Mathematics education and survival with dignity. In H. Alrø, O. R. Christensen, & P. Valero (Eds.), *Critical mathematics education: Past, present and future: Festschrift for Ole Skovsmose* (pp. 51–63). Rotterdam, The Netherlands: Sense Publishers.

Darder, A. (2002). *Reinventing Paulo Freire: A pedagogy of love.* Boulder, CO: Westview.

Darder, A. (2012). *Culture and power in the classroom: Educational foundations for the schooling of bicultural students.* New York, NY: Routledge.

Darder, A., Baltodano, M., & Torres, R. D. (Eds.). (2009). *The critical pedagogy reader* (2nd ed.). New York, NY: Routledge.

Dardot, P., & Laval, C. (2009). *La nouvelle raison du monde: essai sur la société néolibérale* [The novel reasoning of the world: Essays on neoliberal societies]. Paris: Découverte.

De Lissovoy, N., Means, A. J., & Saltman, K. J. (2015). *Toward a new common school movement.* Boulder, CO: Paradigm.

Dewey, J. (1916). *Democracy and education: An introduction to the philosophy of education.* New York, NY: Free Press.

Doganay, A. (2012). A curriculum framework for active democratic citizenship education. In D. Lange & M. Print (Eds.), *Schools, curriculum and civic education for building democratic citizens* (pp. 19–40). Rotterdam, The Netherlands: Sense Publishers.

Ellis, M., & Malloy, C. (2007). *Preparing teachers for democratic mathematics education.* Paper presented at the Mathematics Education in a Global Community Conference, Charlotte, NC.

Ernest, P. (2002a). Empowerment in mathematics education. *Philosophy of Mathematics Education Journal, 15*(1), 1–16. Retrieved from http://people.exeter.ac.uk/PErnest/pome15/ernest_empowerment.pdf

Ernest, P. (2002b). *The philosophy of mathematics education.* New York, NY: Routledge.

Ernest, P. (2002c). *What is empowerment in mathematics education.* Paper presented at the Proceedings of the Third Mathematics Education and Society, Center for Research in Learning Mathematics, Copenhagen.

Ernest, P. (2010). The scope and limits of mathematics education. In H. Alrø, O. R. Christensen, & P. Valero (Eds.), *Critical mathematics education: Past, present and future: Festschrift for Ole Skovsmose* (pp. 65–87). Rotterdam, The Netherlands: Sense Publishers.

Flecha, R. N. (2000). *Sharing words: Theory and practice of dialogic learning.* New York, NY: Rowman & Littlefield Publishers.

Frankenstein, M. (1983). Critical mathematics in education: An application of Paulo Freire's epistemology. *Journal of Education, 165*(4), 315–339.

Frankenstein, M. (1990). Incorporating race, gender, and class issues into a critical mathematical literacy curriculum. *The Journal of Negro Education, 59*(3), 336–347.

Frankenstein, M. (1994). Understanding the politics of mathematical knowledge as an integral part of becoming critically numerate. *Radical Statistics, 56*, 22–40.

Frankenstein, M. (2005). Reading the world with math: Goals for a critical mathematical literacy curriculum. In E. Gutstein & B. Peterson (Eds.), *Rethinking mathematics: Teaching social justice by numbers* (pp. 19–28). Milwaukee, WI: Rethinking Schools.

Frankenstein, M. (2010, June 12). *Critical mathematics education: An application of Paulo Freire's epistemology*. Retrieved from http://webcache.googleusercontent.com/sear ch?q=cache:Q9xVO8G4uskJ:people.exeter.ac.uk/PErnest/pome25/Marilyn Frankenstein Critical Mathematics Education.doc+&cd=1&hl=en&ct=clnk&gl=us

Freire, P. (1998). *Politics and education*. Los Angeles, CA: UCLA Latin American Center Publications.

Freire, P. (2000). *Pedagogy of the oppressed*. New York, NY: Continuum.

Freire, P. (2013). *Education for critical consciousness*. New York, NY: Bloomsbury Academic.

Freire, P., & Faundez, A. (1989). *Learning to question: A pedagogy of liberation*. New York, NY: Continuum.

Freitas, E. (2008). Critical mathematics education: Recognizing the ethical dimension of problem solving. *International Electronic Journal of Mathematics Education Research, 3*(2), 79–93.

Gandin, L. A., & Apple, M. W. (2002). Thin versus thick democracy in education: Porto Alegre and the creation of alternatives to neo-liberalism. *International Studies in Sociology of Education, 12*(2), 99–116.

Giroux, H. (1983). *Theory and resistance in education: Towards a pedagogy for the opposition*. Westport, CT: Bergin & Garvey.

Giroux, H. (1988). *Teachers as intellectuals: Toward a critical pedagogy of learning*. Westport, CT: Bergin & Garvey.

Giroux, H. (1989). *Schooling for democracy: Critical pedagogy in the modern age*. London: Routledge.

Giroux, H. (1994). Toward a pedagogy of critical thinking. In K. Walter (Ed.), *Re-thinking reason: New perspectives in critical thinking* (pp. 199–204). Albany, NY: SUNY Press.

Giroux, H. (2012). *Can democratic education survive in a neoliberal society?* Retrieved May, 2015, from http://truth-out.org/opinion/item/12126-can-democratic-education-survive-in-a-neoliberal-society-startOfPageId12126

Giroux, H. (2014). *Barbarians at the gates: Authoritarianism and the assault on public education*. Retrieved May, 2016, from http://www.truth-out.org/news/item/28272-barbarians-at-the-gates-authoritarianism-and-the-assault-on-public-education

Gramsci, A. (1971). *Prison notebooks: Selections* (G. Nowell-Smith, Ed.). New York, NY: International Publishers.

REFERENCES 159

Groundwater-Smith, S., Brennan, M., McFadden, M., & Mitchell, J. (2003). *Secondary schooling in a changing world.* South Melbourne: Thomson.

Gutstein, E. (2006). *Reading and writing the world with mathematics: Toward a pedagogy for social justice.* New York, NY: Routledge.

Habermas, J. (1972). *Knowledge and human interests.* Boston, MA: Beacon.

Habermas, J. (1975). *Legitimation crisis.* Boston, MA: Beacon.

Habermas, J. (1984). *The theory of communicative action* (M. Thomas, Trans.). Boston, MA: Beacon.

Habermas, J. (1987). *The theory of communicative action* (M. Thomas, Trans.). Boston, MA: Beacon.

Habermas, J. (1990). *Moral consciousness and communicative action.* Cambridge, MA: MIT Press.

Habermas, J. (1996). *Debating the state of philosophy.* Westport, CT: Praeger.

Habermas, J. (2005). *Truth and justification* (B. Fultner, Trans.). Cambridge, MA: MIT Press.

Hannaford, C. (1998). Mathematics teaching is a democratic education. *ZDM, 30*(6), 181–187.

Hayase, N. (2013). *Corporate personhood and the culture of pathology.* Retrieved from http://www.counterpunch.org/2013/02/05/corporate-personhood-and-the-culture-of-pathology/

Hedges, C. (2009). *Empire of illusion: The end of literacy and the triumph of spectacle.* New York, NY: Nation House.

Hedges, C. (2010). *Death of the liberal class.* New York, NY: Nation Books.

Hill, D. (2008). Resisting neo-liberal global capitalism and its depredations: Education for a new democracy. In D. E. Lund & P. R. Carr (Eds.), *Doing democracy: Striving for political literacy and social justice* (pp. 33–49). New York, NY: Peter Lang.

Hill, D. (2009). *Contesting neoliberal education: Public resistance and collective advance.* New York, NY: Routledge.

hooks, b. (2003). *Teaching community: A pedagogy of hope.* New York, NY: Routledge.

Horn, I. S. (2014). *Strength in numbers: Collaborative learning in secondary mathematics.* Reston, VA: National Council of Teachers of Mathematics.

Hursh, D. (2007a). Assessing no child left behind and the rise of neoliberal education policies. *American Educational Research Journal, 44*(3), 493–518.

Hursh, D. (2007b). Marketing education: The rise of standardized testing, accountability, competition, and markets in public education. In E. W. Ross & R. Gibson (Eds.), *Neoliberalism and education reform* (pp. 15–34). Cresskill, NJ: Hampton Press.

Hyslop-Margison, E. J., & Naseem, M. A. (2007). *Scientism and education.* New York, NY: Springer.

Hyslop-Margison, E. J., & Thayer, J. (2009). *Teaching democracy: Citizenship education as critical pedagogy.* Rotterdam, The Netherlands: Sense Publishers.

Kemmis, S. (2008). Critical theory and participatory action research. In P. Reason & H. Bradbury (Eds.), *The Sage handbook of action research: Participative inquiry and practice* (pp. 121–138). London: Sage Publications.

Kemmis, S., McTaggart, R., & Nixon, R. (2014). *The action research planner*. New York, NY: Springer.

Kennedy, N. S. (2009). Towards a dialogical pedagogy: Some characteristics of a community of mathematical inquiry. *Eurasia International Journal of Mathematics, Science, and Technology Education, 5*(1), 71–78.

Kohn, A. (1992). *No contest: The case against competition*. Boston, MA: Houghton Mifflin.

Kohn, A. (1999). *The schools our children deserve: Moving beyond traditional classrooms and "tougher standards."* Boston, MA: Houghton Mifflin.

Kohn, A. (2000). *The case against standardized testing: Raising the scores, ruining the schools*. Portsmouth, NH: Heinemann.

Kohn, A. (2004). *What does it mean to be well educated?* Boston, MA: Beacon.

Kohn, A. (2006). *Beyond discipline from compliance to community*. Alexandria, VA: Association for Supervision and Curriculum Development.

Kohn, A., & Shannon, P. (Eds.). (2002). *Education, Inc.: Turning learning into a business*. Portsmouth, NH: Heinemann.

Kovacs, P., & Christie, H. (2008). The Gates' foundation and the future of U.S. public education. *Journal for Critical Education Policy Studies, 6*(2), 1–15.

Kozol, J. (2005). *The shame of the nation: The restoration of apartheid schooling in America*. New York, NY: Crown Publishers.

Leistyna, P. (2007). Neoliberal nonsense. In P. K. McLaren & J. L Kincheloe (Ed.), *Critical pedagogy: Where are we now?* New York, NY: Peter Lang.

Lincove, J. A. (2009). Are markets good for girls? The world bank and neoliberal education reforms in developing countries. *Journal of Diplomacy and International Relations, 10*, 59. Retrieved from http://www.utexas.edu/lbj/chasp/research/downloads/Doc8Whitehead2009.pdf

Lund, D., & Carr, P. (2008). Scanning democracy. In D. Lund & P. Carr (Eds.), *Doing democracy: Striving for political literacy and social justice* (pp. 1–29). New York, NY: Peter Lang.

Manconi, L., Aulls, M. W., & Shore, B. M. (2008). Teachers' use and understanding of strategy in inquiry instruction. In M. A. Delcourt, M. W. Aulls, & B. M. Shore (Eds.), *Inquiry in education* (Vol. 2, pp. 247–270). New York, NY: Lawrence Erlbaum Associates.

Marcuse, H. (2013). *One-dimensional man: Studies in the ideology of advanced industrial society*. London: Routledge.

McLaren, P., & Kincheloe, J. L. (Eds.). (2007). *Critical pedagogy: Where are we now?* New York, NY: Peter Lang.

REFERENCES 161

McLaren, P., & Leonard, P. (1993). *Paulo Freire: A critical encounter.* New York, NY: Routledge.

McNeil, L. (2009). Standardization, defensive teaching, and the problems of control. In A. Darder, M. Baltodano, & R. D. Torres (Eds.), *The critical pedagogy reader* (pp. 384–396). New York, NY: Routledge.

Morrow, R. A., & Torres, C. A. (2002). *Reading Freire and Habermas: Critical pedagogy and transformative social change.* New York, NY: Teachers College Press.

Mortimer, E., & Scott, P. (2003). *Meaning making in secondary science classrooms.* Buckingham: Oxford University Press.

Murphy, M., & Fleming, T. (2010). *Habermas, critical theory and education.* New York, NY: Routledge.

Nagda, B. A., Gurin, P., & Lopez, G. E. (2003). Transformative pedagogy for democracy and social justice. *Race, Ethnicity and Education, 6*(2), 165–191.

National Council of Teachers of Mathematics. (2000). *Principles and standards for school mathematics.* Reston, VA: National Council of Teachers of Mathematics.

Nicholas, B., & Bertram, B. (2001). Theory and research on teaching as dialogue. In V. Richardson (Ed.), *Handbook of research on teaching* (pp. 1102–1121). Washington, DC: American Educational Research Association.

Noddings, N. (2003). *Happiness and education.* Cambridge: Cambridge University Press.

Nystrand, M. (1997). *Opening dialogue: Understanding the dynamics of language and learning in the English classroom.* New York, NY: Teachers College Press.

Orlowski, P. (2012). *Teaching about hegemony: Race, class and democracy in the 21st century.* New York, NY: Springer.

Pietsch, J. (2009). *Teaching and learning mathematics together: Bringing collaboration to the centre of the mathematics classroom.* Newcastle: Cambridge Scholars Publishing.

Pine, G. J. (2009). *Teacher action research: Building knowledge democracies.* Los Angeles, CA: Sage Publications.

Powell, A. B., & Frankenstein, M. (1997). *Ethnomathematics: Challenging eurocentrism in mathematics education.* Albany, NY: SUNY Press.

Ravitch, D. (2011). *The death and life of the great American school system: How testing and choice are undermining education.* New York, NY: Basic Books.

Ravitch, D. (2016). When public goes private, as Trump wants: What happens. *The New York Review of Books, 63*(19), 58–61.

Rogers, C. (1995). *A way of being.* New York, NY: Houghton Mifflin.

Rogers, C., & Freiberg, J. (1994). *Freedom to learn.* New York, NY: Merrill.

Ross, E. W., & Gibson, R. J. (2007). *Neoliberalism and education reform.* Cresskill, NJ: Hampton Press.

Sacks, P. (2009). *Standardized minds: The high price of America's testing culture and what we can do to change it.* Cambridge, MA: Da Capo Press.

Schneider, M. K. (2015). *Common core dilemma—who owns our schools?* New York, NY: Teachers College Press.

Schneider, M. K. (2016). *School choice: The end of public education*. New York, NY: Teachers College Press.

Scott, T. (2011). A nation at risk to win the future: The state of public education in the U.S. *Journal for Critical Education Policy Studies, 9*(1), 268–316. Retrieved from http://www.jceps.com/archives/666

Shor, I. (1987). *Critical teaching and everyday life*. Chicago, IL: University of Chicago Press.

Shor, I. (1993). Education is politics. In P. McLaren & P. Leonard (Eds.), *Paulo Freire: A critical encounter* (pp. 25–35). New York, NY: Routledge.

Simmt, E. (2001). Citizenship education in the context of school mathematics. *Canadian Social Studies Journal, 35*(3). Retrieved from http://www.educ.ualberta.ca/css/Css_35_3/ARcitizenship_education.htm

Skovsmose, O. (1994). *Towards a philosophy of critical mathematics education*. London: Kluwer Academic Publishers.

Skovsmose, O. (2011). *An invitation to critical mathematics education*. Rotterdam, The Netherlands: Sense Publishers.

Skovsmose, O., & Alrø, H. (2004). *Dialogue and learning in mathematics education: Intention, reflection, critique* (Vol. 29). New York, NY: Springer.

Skovsmose, O., & Borba, M. (2000). *Research methodology and critical mathematics education*. Denmark: Centre for Research in Learning Mathematics.

Skovsmose, O., & Greer, B. (2012). *Opening the cage: Critique and politics of mathematics education*. Rotterdam, The Netherlands: Sense Publishers.

Smyth, J. (2011). *Critical pedagogy for social justice*. London: Continuum.

Spring, J. (2014). *How educational ideologies are shaping global society: Intergovernmental organizations, NGOs, and the decline of the nation-state*. New York, NY: Routledge.

Staples, M. (2007). Supporting whole-class collaborative learning in a secondary mathematics classroom. *Cognition and Instruction, 25*(2), 1–57.

Stevenson, N. (2011). *Education and cultural citizenship*. London: Sage Publications.

Torbat, A. E. (2008). *Global financial meltdown and the demise of neoliberalism*. Retrieved from http://www.jonbesh-iran.com/Jonbesh/Site/English/2008/Oktober/tarjomeh242%5B1%5D.pdf

Tsankova, J., & Dobrynina, G. (2005). Developing curious students. In R. Audet & L. Jordan (Eds.), *Integrating inquiry across the curriculum* (pp. 85–109). Thousand Oaks, CA: Corwin Press.

Valero, P., & Zevenbergen, R. (2004). *Researching the socio-political dimensions of mathematics education: Issues of power in theory and methodology* (Vol. 35). New York, NY: Springer.

Ventura, P. (2012). *Neoliberal culture: Living with American neoliberalism*. Burlington: Ashgate Publishing.

REFERENCES

Vygotsky, L. S. (1978). *Mind in society: The development of higher psychological processes*. Cambridge, MA: Harvard University Press.

Wells, G. (1999). *Dialogic inquiry: Towards a sociocultural practice and theory of education*. New York, NY: Cambridge University Press.

Wells, G. (2009). Dialogic inquiry as collaborative action research. In S. Noffke & B. Somekh (Eds.), *The Sage handbook of educational action research* (pp. 50–61). London: Sage Publications.

West, C., & Smiley, T. (2012). *The rich and the rest of us*. New York, NY: Hay House.

Westheimer, J. (2015). *What kind of citizen? Educating our children for the common good*. New York, NY: Teachers College Press.

Index

Academic standards 104
Accountability system 27
Action research 4, 44, 113, 156, 160, 164
Actions
  critical participatory 164
  transformative 109
  voluntary 119
Activities
  collaborative 62, 91
  context-dependent 26
  non-graded 9
  subversive 14
  test-preparation 3, 26
Administrative sanction 56
African American students 125
Agencies 34, 164
  critical 24
Agenda 1, 36, 144, 156, 164
  corporate 1, 8, 23, 30
  scripted 144
  secret 149
Agentic participants 135, 147
Alienation 96
American neoliberalism 162
Anti-dialogical 95, 110
Anti-intellectualism 27
  anti-racist 50
Atmosphere 128
  anti-dialogic 112
  welcoming 42
Authoritarian ix, 88, 95, 96, 104, 148, 149
  anti-intellectualism 164
  approach 95, 96, 114, 148
  education 115
  teaching 148
Authority-dependent personalities 95

Banking education 108
Black community 50
  poor working-class 50
Business mentality 25

Capitalism 24, 164
  counter neoliberal 51
  neo-liberal 50
Capitalist ruling classes 24, 34
  chauvinism 164

Caring 73–75, 79–81, 120, 139
  caring individuals 25, 28, 29
Citizens 1, 2, 16, 21, 29, 42, 46, 79, 87,
  100–102, 116, 141, 152, 163, 164
  critical xii, 6, 23, 25, 29, 30, 37, 44,
    45, 53, 84, 87, 89, 95, 126, 134,
    140–142
  democratic 142, 156, 157
  passive-obedient 141
Citizenship viii, ix, xi, 8, 29, 65, 134, 135, 137,
  139–143, 145, 165
  consumer-based neoliberal 139
  individual 143
  instrumental 140
  market-based 135
  social justice-based 2, 74
  citizenship and mathematics
    education 149, 164
Class consciousness x, 5, 6, 37, 84
And mathematical literacy 37, 39, 41, 43, 45,
  47, 49, 51
Classroom-based research 1, 3, 35, 144, 146,
  149, 151
Classroom life 149
  colonize 152
  democratize 2, 136, 152
Collaboration 6, 12–14, 42, 44, 53, 54, 58,
  61–63, 65, 78, 81, 83, 85, 95, 118–120,
  126, 127
  equal 118
  genuine 28
  inquiry-based 70, 89, 132
Colonized life-world 51, 98, 106, 122
Communicative rationality 108–110, 118, 126,
  138, 146, 150, 151, 164
Community xi, xii, 55, 58, 67, 68, 73–75, 79,
  80, 82, 83, 85, 114–117, 125, 127, 142, 143,
  145, 147, 148, 160
  collaborative 123
  democratic xi, 125, 135, 152
  discursive 110
  inclusive 136
  local 68
  micro 126
  underserved 61
Community service 5, 67, 72, 80, 84, 88,
  137, 141

INDEX

Competitive learning 53, 55, 57–59, 61, 63, 65, 72, 76, 85, 118, 119, 121, 147
  individualistic 147
  democratic 155
Conservative restoration process 33
Consumer-based citizenship 139
Contesting neoliberal education 159
Counterhegemonic learning and teaching processes 84
Critical citizenship xi, xii, 1, 2, 4, 6, 30, 35, 45, 118, 120, 133, 135, 137, 140–143, 145, 146, 149
Critical mathematics education (CME) 1–4, 6, 35, 52–55, 104, 105, 107, 111, 113–115, 117–119, 125–131, 133, 134, 143–148, 150–153, 157, 158, 162
Critical mathematical literacy 5, 6, 43, 45–48, 74, 95, 104, 113, 129, 134, 137, 138, 146, 149, 152, 164
Critical participatory action research (CPAR) 1, 3, 4, 8, 9, 83, 134, 146, 152
Critical pedagogy (CP) 1, 3, 4, 13, 34, 35, 37, 49, 63, 76, 84–86, 98, 102, 121–123, 144, 145, 158–160, 164
Critical theory and participatory action research 160

Dehumanizing education 124
Democratic citizenship xi, 30, 139, 164
  active 157
  thick 136
Demoralized and de-professionalized teachers 29
Dialogic classroom 14, 17, 64, 88, 94, 97, 107, 113, 115, 124, 135, 139
Dialogic pedagogy 2, 5, 6, 69, 87, 88, 96, 104, 105, 107, 116, 118, 127, 134, 137, 146–50, 152

Edu-business 6, 34, 164
Egalitarian collaboration 55, 97, 118, 120, 124, 129, 133, 135, 136
Elite 164
  corporate 24
  ruling 50
Emancipatory praxis 138, 164
Empathy 50, 78, 80, 86, 99, 112, 115, 117, 119, 124, 125, 149, 164
Ethnic pride 13, 50
Ethnomathematics 54, 161

Exercise paradigm 95, 109, 147, 150

Facilitative classroom environment 148
Frankfurt school 141
Functional literacy 17, 38, 47, 74, 108–110, 118, 129, 139, 143, 164

Group work and whole-class Discussion 116, 117
Growth 135
  communal 135
  economic 30
  individual 135

Habermas xi, 69, 78, 85, 86, 88, 103, 106, 108–111, 117, 137, 141, 142, 148, 150, 159, 161
Hegemony 33, 49, 69, 76, 88, 103, 146, 161
Humanizing education 4, 15, 40, 69, 108, 109, 112, 152

Ideal speech situation 1, 78, 85, 86, 117, 118, 137, 148, 149
Inequalities viii, 28, 37, 67, 69, 71–73, 84, 120, 124, 127, 128
  educational 155
  linear 67, 71, 78, 125
Inquiry viii, xi, 4, 10, 14, 71, 85, 110, 111, 117, 118, 126–133, 135, 140, 146, 147, 155, 156, 160
Instrumental rationality 108–110, 126, 138

Justice-oriented citizenship 139

Knowledge xi, xii, 3, 5, 6, 8, 10, 16–18, 20, 21, 72–74, 85–87, 96, 108–110, 115, 128, 135, 142, 143
  co-constructed 95
  common xi
  conceptual 9
  pre-packaged 25, 108
  transmitted 112

Labor 12, 29, 50, 129
  employee's 138
Learning process 4, 6, 12, 41, 64, 72, 104, 121, 134, 136, 146
  collective 12, 40
  facilitative 147
  inquiry-based 46
  non-hierarchical 40

INDEX

Life-world   9, 37, 38, 41, 45, 49, 52, 64–66, 87, 90, 101, 103, 106, 111, 117, 118, 129–131
Literacy   45, 74, 157, 159, 164
   critical   xi, 16, 17, 38, 44–47, 51, 74, 105, 108, 109, 117, 128–131, 133, 139, 140, 149
   political   159, 160

Mathematical literacy   37, 39, 41, 43, 45
   and citizenship   137, 165

Neoconservatives   33, 34, 149
   neoliberal   viii, x, 2, 4, 23, 24, 26, 29, 30, 32–35, 55, 56, 106, 131, 132, 137, 138, 143–145, 149, 152
Neoliberal pedagogy   4, 6, 12, 35, 53, 73, 84, 120–122, 134, 135, 138, 147, 149, 151, 153
   consumer-based   139, 164
   hegemony of   6, 131

Open-ended word problems   37, 89, 109, 110, 147, 165
Oppression   4, 40, 100, 138, 147

Participation   xi, 8, 9, 23, 40, 41, 64, 111, 119, 122, 127, 132, 135, 139, 140
   active   9, 55, 90, 95, 96, 121
   voluntary   8, 115, 120
Participatory citizenship   80, 165
   justice-based   134
   responsible   139
Paulo Freire   1, 161, 162
Pedagogy   xi, xii, 1, 12, 25, 26, 31, 34, 50, 51, 107, 108, 118, 129, 137, 140, 142, 143, 153, 157–159
   dialogical   160
   emerging   155
   oppressive   108
   traditional   18, 19
Positivism   140, 141
Power relations and dialogue   111
Praxis   xii, 3, 69, 118, 121, 136
   continual   23
   educational   69

Quadratic function   40, 41
Quantified measurement   28
Question authorities   150

Race   6, 13, 27, 28, 49, 50, 59, 91, 123, 125, 161
Race theory   50
Race to the Top (RTT)   27, 31
Reinventing Paulo Freire   157
Revolutionary changes   51
Romantic optimism   62
Ruling class   25, 29, 50, 51, 84

Safety net   78, 79, 87
Scripted curriculum   130
Self-assigned leaders   52, 85, 122, 124
Selfishness   67, 69, 80, 81, 84
Skill-drill examples   95
Society   28, 31, 32, 46–48, 51, 63, 78–82, 84, 87, 98, 100, 103, 104, 129, 130, 139–142, 149, 150, 156, 157
   advanced industrial   160
   class-based   6, 37, 49, 51, 138
   classless   46
   collaborative   75
   competitive   75, 81
   democratic   30, 116
   equal   79
   free   34
   global   162
   micro   116, 136, 147
   winner-takes-all   81
Socioeconomic class   28
Solidarity, ix   5, 6, 12, 28, 50, 65, 67, 68, 73, 81, 82, 84, 87, 139, 151, 165
   peer   41
   social   84
Standardization movement, ix   3, 23, 25–27, 34, 144
   market-based   146
STEM   30
Student Loan Crises   88, 89, 91, 93, 95, 97, 99, 101, 103, 105
Subversive teaching practices   51

Teacher accountability   x, 27, 28
Thematic discussion   98
Theory of communicative action   159, 165
Transformative social changes   35, 161, 165

Unemployment   68, 120
Unethical assumption   27

INDEX

Values   19, 23, 31, 53, 61–63, 69, 73, 74, 81, 86, 87, 116, 120–122, 124, 134–136, 139, 142
  common   61
  established   132
  ethical   136
  hegemonic   87
  universal   60, 62, 63, 65
Vygotsky   13, 41, 71, 78, 85, 123, 148, 163

White working-class students   50
Whole-class discussion   5, 10, 11, 13, 14, 17–19, 39, 43, 47, 53, 57, 97, 111, 117, 118, 121, 124, 130
Working-class black people   51
Working-class students   31, 51
World citizens   137

ZPD (Zone of Proximal Development)   12, 13, 41, 42, 71, 72, 85, 91, 123, 124, 127, 148

Printed in the United States
By Bookmasters